Air Transport and the Environment

T0225679

BEN DALEY
Centre for Development, Environment and Policy
School of Oriental and African Studies
University of London

Routledge
Taylor & Francis Group

LONDON AND NEW YORK

First published in paperback 2024

First published 2010 by Ashgate Publishing

Published 2016 by Routledge
4 Park Square, Milton Park, Abingdon, Oxon OX14 4RN

and by Routledge
605 Third Avenue, New York, NY 10158

Routledge is an imprint of the Taylor & Francis Group, an informa business

Publisher's Note
The publisher has gone to great lengths to ensure the quality of this reprint but points out that some imperfections in the original copies may be apparent.

British Library Cataloguing in Publication Data
Daley, Ben.
 Air transport and the environment.
 1. Aeronautics, Commercial--Environmental aspects.
 2. Aircraft exhaust emissions--Environmental aspects.
 3. Aeronautics and state.
 I. Title
 387.7-dc22

Library of Congress Cataloging-in-Publication Data
Daley, Ben.
 Air transport and the environment / by Ben Daley.
  p. cm.
 Includes bibliographical references and index.
 ISBN 978-0-7546-7286-9 (hardback) -- ISBN 978-0-7546-9089-4
 (ebook) 1. Aeronautics--Environmental aspects. I. Title.
 TD195.A27D35 2010
 387.7--dc22

                                                                2010015516

ISBN: 978-0-7546-7286-9 (hbk)
ISBN: 978-1-03-283853-3 (pbk)
ISBN: 978-1-315-56632-0 (ebk)

DOI: 10.4324/9781315566320

AIR TRANSPORT AND THE ENVIRONMENT

Contents

List of Figures   *vii*
List of Tables   *ix*
List of Abbreviations   *xi*
Preface   *xv*

1   Introduction   1

2   Understanding the Issues   19

3   Air Transport and Climate Change   49

4   Air Transport and Air Quality   87

5   Aircraft Noise   123

6   Air Transport and Sustainable Development   165

7   Conclusion   203

References   *213*
Index   *245*

List of Figures

2.1	Aircraft engine emissions under ideal and actual conditions	22
2.2	Kerosene combustion products under actual conditions	22
2.3	Global CO_2 emissions from aviation	25
2.4	Global CO_2 emissions from aviation under various scenarios	25
2.5	Improvements in aircraft fuel efficiency	35
2.6	The growth of air transport expressed in terms of passenger numbers	39
6.1	Sustainable development: linking economy, society and environment	168
6.2	Some objectives of sustainable development	168

List of Tables

1.1	The main environmental impacts of air transport	5
1.2	ICAO inventory of aviation environmental problems	14
2.1	Aircraft emissions under different engine operating regimes	23
3.1	The main effects of aircraft on climate	57
4.1	The main effects of aircraft on air quality	95
5.1	Some measures of aircraft noise	128

List of Abbreviations

ACARE	Advisory Council for Aeronautics Research in Europe
AEI	Average exposure indicator
AERO	Aviation Emissions and Evaluation of Reduction Options
AETIAQ	Aviation Emissions and Their Impact on Air Quality
AMAN	Arrival management
ANASE	Attitudes to Noise from Aviation Sources in England
ANCAT	Abatement of Nuisances Caused by Air Transport
ANIS	Aircraft Noise Index Study
APD	Air Passenger Duty
APU	Auxiliary power unit
AQMA	Air Quality Management Area
AQS	Air Quality Strategy
ASA	Air service agreement
ATAG	Air Transport Action Group
ATC	Air Traffic Control
ATM	Air Traffic Management
CAA	Civil Aviation Authority
CAEE	Committee on Aircraft Engine Emissions
CAEP	Committee on Aviation Environmental Protection
CAN	Committee on Aircraft Noise
CATE	Centre for Air Transport and the Environment
CDA	Continuous descent approach
CDM	Collaborative decision-making
CEC	Commission of the European Communities
CeDEP	Centre for Development, Environment and Policy
CFD	Computational fluid dynamics
CH_4	Methane
CIs	Chemi-ions
CNEL	Community noise equivalent level
CNS	Communications, navigation and surveillance
CO	Carbon monoxide
CO_2	Carbon dioxide
CSP	The Chartered Society of Physiotherapy
DEFRA	Department for Environment, Food and Rural Affairs
DESA	Department of Economic and Social Affairs

DETR	Department for Environment, Transport and the Regions
DfID	Department for International Development
DfT	Department for Transport
DLR	Deutsches Zentrum für Luft-und Raumfahrt
DMAN	Departure management
DNL	Day-night average sound level
DOTARS	Department of Transport and Regional Services
DTI	Department of Trade and Industry
EC	European Commission
ECAC	European Civil Aviation Conference
EDF	Environmental Defense Fund
EEA	European Environment Agency
EEG	Electroencephalograph
EIA	Environmental Impact Assessment
EIDG	Emissions Inventory Database Group
EPA	Environmental Protection Agency
ETS	Emissions Trading Scheme
FAA	Federal Aviation Administration
FDI	Foreign direct investment
FEGP	Fixed electrical ground power
FESG	Forecasting and Economic Sub-Group
FTK	Freight tonne-kilometre
GDP	Gross domestic product
GIS	Geographic Information System
GSG	Global Scenario Group
GWP	Global warming potential
H_2	Hydrogen
H_2O	Water
H_2SO_4	Sulphuric acid
HCs	Hydrocarbons
HFCs	Hydrofluorocarbons
HNO_3	Nitric acid
HONO	Nitrous acid
IATA	International Air Transport Association
ICAO	International Civil Aviation Organization
INM	Integrated Noise Model
IPCC	Intergovernmental Panel on Climate Change
IPPR	Institute for Public Policy Research
ITA	International tourist arrival
LCC	Low-cost carrier
Ldn	Day-night average sound level

L_{den}	Day-evening-night average sound level
L_{eq}	Equivalent continuous sound level
LIDAR	Light Detection and Ranging
L_{max}	Instantaneous maximum sound level
LNG	Liquefied natural gas
LP/LD	Low-power, low-drag
LPG	Liquefied petroleum gas
LTO	Landing and take-off (cycle)
MDG	Millennium Development Goal
MIT	Massachusetts Institute of Technology
N_2	Nitrogen
N_2O	Nitrous oxide
NAAQS	National Ambient Air-Quality Standards
NADP	Noise abatement departure procedure
NAFC	North Atlantic Flight Corridor
NGO	Non-governmental organisation
NO	Nitric oxide
NO_2	Nitrogen dioxide
NO_x	Nitrogen oxides
NPL	National Physics Laboratory
NPR	Noise preferential route
O_2	Oxygen
O_3	Ozone
OEF	Oxford Economic Forecasting
OH	Hydroxyl
PFCs	Perfluorocarbons
PM	Particulate matter
ppb	parts per billion
ppmv	parts per million by volume
QC	Quota Count
RCEP	Royal Commission on Environmental Pollution
RFI	Radiative forcing index
RPK	Revenue passenger-kilometre
RTK	Revenue tonne-kilometre
SAGE	System for assessing Aviation's Global Emissions
SARS	Severe acute respiratory syndrome
SBSTA	Subsidiary Body on Scientific and Technical Advice
SDC	Sustainable Development Commission
SEL	Sound exposure level
SF_6	Sulphur hexafluoride
SIP	State Implementation Plan

SO$_2$	Sulphur dioxide
SO$_3$	Sulphur trioxide
SO$_x$	Sulphur oxides
SRES	Special Report on Emissions Scenarios
T&E	European Federation for Transport and Environment
UHCs	Unburned hydrocarbons
UN	United Nations
UNCED	United Nations Conference on Environment and Development
UNCHE	United Nations Conference on the Human Environment
UNCSD	United Nations Commission on Sustainable Development
UNCTAD	United Nations Conference on Trade and Development
UNDP	United Nations Development Programme
UNFCCC	United Nations Framework Convention on Climate Change
UNWTO	United Nations World Tourism Organization
VAT	Value Added Tax
VOC	Volatile organic compound(s)
WCED	World Commission on Environment and Development
WHO	World Health Organization
WRI	World Resources Institute
WSSD	World Summit on Sustainable Development
WWF	World Wide Fund for Nature

Preface

Air transport is a popular, rapidly-growing industry that provides a wide range of important economic and social benefits. Those benefits include employment, trade, tourism, investment, enhanced productivity, increased competitiveness, knowledge transfer, greater mobility and many multiplier effects. Yet there are now unprecedented levels of popular and scientific concern about the environmental impacts of flying. Emissions from aircraft and airports contribute to climate change and to localised air pollution, and aircraft noise affects individuals in the vicinity of airports in a variety of ways. Other environmental impacts associated with airports include water consumption, water pollution, land contamination and ecological effects on species and their habitats. Concerns about some aviation environmental impacts are already acute and are likely to become yet more prominent as the demand for air transport grows. Reconciling such conflicting concerns presents policymakers with a major challenge – especially given that, in many cases, air transport operates across national borders and connects individuals and communities that may have become highly dependent on air transport for their livelihoods.

This book provides an overview of the main issues relating to the environmental impacts of air transport. It explains the central challenge facing policymakers in terms of sustainable development, focusing on the importance of balancing the industry's economic, social and environmental costs and benefits, both for people living now and for future generations. In individual chapters, current scientific knowledge of the main aviation environmental impacts – climate change, local air pollution and aircraft noise – is presented. Various responses to those issues are also considered, including a range of policy options based on regulatory, market-based and voluntary approaches. Key concepts such as environmental capacity, radiative forcing and carbon offsetting are explained. In addition, this account emphasises the critical implications of aviation environmental issues for policymakers and for the management of the air transport industry.

This book has been written for professionals in the air transport industry, policymakers and regulators. It is also intended for use by academic researchers, students and others who are interested in the broader aspects of the relationship between air transport and the environment. This book is based on an extensive review of the academic literature and on primary research into aviation environmental impacts undertaken between 2006 and 2008 with the Centre for Air Transport and the Environment (CATE) at the Manchester Metropolitan University, and subsequently at the Centre for Development, Environment and Policy (CeDEP) at the School of Oriental and African Studies, University of London. I am grateful to the many colleagues in both institutions – and also to a wide range of professionals in the air transport industry – who have provided much-needed advice and assistance during this research.

Debates about the environmental impacts of flying often generate strongly polarised reactions, yet this book adopts a constructive approach to the subject and attempts to present the environmental issues in a clear, straightforward manner. Its central purpose is to bring the sustainable development challenge facing the air transport industry to the fore – and so to inform effective policy responses. Air transport plays a vital role in supporting economies and societies that are increasingly interconnected by globalisation. This book presents the view that the vital economic and social benefits of the air transport industry should not be lost – and in fact could be distributed far more widely and equitably – but that the environmental impacts of air transport nevertheless require urgent and effective management.

Ben Daley
Centre for Development, Environment and Policy
School of Oriental and African Studies
University of London

Reviews for
Air Transport and the Environment

This is a thorough and very readable review of the environmental and social impacts of aviation, as both the industry and policymakers struggle to reconcile aviation growth with environmental limits. For those wanting a concise but detailed understanding of the issues and literature, this book is a very good place to start.

Paul Upham, University of Manchester, UK

Given the highly charged debates surrounding aviation's environmental impacts, and their consequences for public policy, no publication currently available covers the issues adequately. Ben Daley's book more than fills this gap. In lucid and uncompromising terms, he spells out the impacts objectively and convincingly, with climate change, air quality, noise and sustainable development receiving extended treatment. This is thought provoking and challenging material that demands our attention.

David Gillingwater, Loughborough University, UK

His explanations of each key area lay the groundwork for detailed discussions of the problems and policy options facing the industry and its regulators.

Dawna L. Rhoades, Embry-Riddle Aeronautical University, USA

Ben Daley's book provides a comprehensive, readable, and balanced introduction to the complex field of aviation and the environment. It argues that swift and radical change will be needed to achieve sustainable development in the sector, for which trade-offs between environmental and economic concerns will have to be accepted. The book deserves to be widely read by policy-makers and all those interested in measurable progress on these matters.

Stefan Gössling, Linnaeus University, Sweden

Sustainable development is a formidable challenge for aviation. Air Transport and the Environment *presents the scope of this challenge within a scientifically-comprehensive, critical and thought-provoking analysis. By drawing together detail on pollution, noise and climate change whilst introducing the broader policy landscape, Daley is well placed to delve into arguably more contentious issues surrounding sustainable development. His final chapter questioning aviation's economic contribution, in light of the unequal benefits accrued, provides an insightful addition to this book and the wider debate.*

Alice Bows, Sustainable Consumption Institute,
University of Manchester, UK

1 Introduction

A Critical Challenge

Air transport is associated with a wide variety of ideas, representations and meanings. Flying, perhaps more than any other means of transport, conjures diverse images of leisure, recreation, connectivity, business productivity, social contact and cultural exchange. No other mode of transport offers such ease of access to international tourism destinations, such rapid transfer between major cities at the continental or global scales, or such a wide range of opportunities for travellers to encounter and experience new places. Given the capacity of the industry to drive economic growth and to enhance mobility, air transport has become an important instrument of globalisation (defined broadly as the process of increasing integration of economic, political, social and cultural activities at the global scale) and a potent symbol of those new patterns of interconnection. Yet, to some people, air transport represents a destructive force that threatens the integrity of communities and environments. Despite the emergence of the so-called low-cost carriers (LCCs), flying is still regarded by many people as a luxury enjoyed by affluent travellers with lifestyles of unprecedented leisure. Many others do not enjoy the benefits of aviation but nonetheless suffer its impacts: noise, pollution, congestion and – sometimes – dislocation. To some people, air transport represents an industry that has enjoyed unfair subsidies and that has been regulated with an excessively light touch. Air transport therefore generates fierce controversy, as in debates about the construction of new airport infrastructure, about the social acceptability of night flying or about the supply of air-freighted organic food. Above all, aviation has become a powerful symbol of fossil fuel consumption, greenhouse gas emissions and climate change. Such diverse images illustrate the complexity and relevance of the relationship between air transport and the environment.

As with many other environmental issues, debates about the environmental impacts of air transport are often framed within wider discussions of sustainable development. Put crudely, sustainable development means balancing the economic, social and environmental benefits and costs of development, both for people living now and for future generations (Adams 2009; Baker 2006; Dresner 2008; Elliott 2006; WCED 1987). Air transport is frequently celebrated because it provides important economic and social benefits. The economic benefits of air transport have been well documented, at least at the national scale (Boon and Wit 2005; DfT 2003a; OEF 1999; 2002; 2006). Aviation makes a vital contribution to economic development with the result that the industry is frequently described as a major engine of economic growth. That economic contribution occurs primarily through the influence of aviation on the performance of other industries and by supporting their growth: air transport facilitates greater access to markets, specialisation, economies of scale and foreign direct investment (FDI). In addition, however, air transport makes a substantial contribution to local, regional and national economies in its own right: by adding to gross domestic product (GDP), by creating direct and indirect employment, by raising productivity, by exporting goods and services, by contributing taxes and through investment. Thus aviation is regarded as a key component of transport infrastructure on which many other parts of national economies depend, and 'investments in that infrastructure boost productivity growth across the rest of the economy' (OEF 1999, 5). Aviation allows markets to expand – potentially to the global scale – and stimulates technological improvements and innovation. Therefore air transport is both an important economic sector in its own right and a vital facilitator of growth in other sectors. In addition to its economic benefits, air transport provides many social benefits associated with employment, leisure, recreation, cultural exchange, educational opportunities and greater access to family and friends (Bishop and Grayling 2003; Caves 2003; Shaw and Thomas 2006).

Given the importance of air transport for economic and social development – and the broad popularity of air travel – the aviation industry has generally sustained high rates of growth (of around five per cent per year) over the last five decades. During that period, aviation growth rates have tended to exceed the rate of global economic growth (Bailey 2007; Humphreys 2003; IPCC 1999; Lee 2004; Lee et al. 2009). In particular, the sustained growth of demand for air transport has been reinforced by two important trends: globalisation and the growth of tourism (Debbage 1994; Goetz and Graham 2004; Janelle and Beuthe 1997; Mayor and Tol 2010; Pels 2008). Globalisation is now acknowledged to be a complex process that is responsible for profound economic, political, social, cultural and environmental transformations worldwide. Communications between many places are now almost instantaneous; intercontinental transportation is commonplace; the world economy has become increasingly integrated;

the influence of multinational organisations has expanded; and the autonomy of most nation-states has diminished (Hettne 2008). Air transport has been important in facilitating the process of globalisation, although globalisation has in turn increased the demand for air travel (Goetz and Graham 2004; Janelle and Beuthe 1997; Young 1997). The growth of demand for air transport has also been closely linked to the rising demand for tourism, an industry that is highly dependent upon the availability of rapid, long-distance air services (Bieger and Wittmer 2006; Daley et al. 2008; Gössling and Peeters 2007; Graham et al. 2008; May 2002; Mayor and Tol 2010). At the global scale, tourism, like aviation, is an important economic driver and is projected to expand rapidly, at an average rate of around four per cent per year, until at least 2020. Projections by the United Nations World Tourism Organization (UNWTO) have indicated that international tourist arrivals (ITAs) will double between 2005 and 2020 and are expected to reach 1.6 billion by the latter year (UNWTO 2007). Mayor and Tol (2010) have argued that international tourism is projected to grow substantially over the decadal timescale, with increases in the number and length of trips, and with growth being driven largely by increasing demand by consumers in Asian countries. All of these factors combine to suggest that strong, sustained growth in demand for air transport is likely to continue beyond 2030 (Bieger et al. 2007; Bows et al. 2005; 2006; DfT 2003c; IPCC 1999; Lee et al. 2009). By 2050, air passenger traffic is expected to have increased five-fold from 1995 levels (IPCC 1999).

Yet in addition to its substantial economic and social benefits, air transport has a range of significant – and generally increasing – environmental impacts (Table 1.1). Some aviation environmental impacts are local to airports, whilst others are of global concern. Aircraft noise has long been a source of nuisance for individuals living in the vicinity of airports and beneath their arrival and departure routes. Aircraft and airports also generate air pollution, especially due to the emission of nitrogen oxides (NO_x) and particles. Airports also have other localised environmental impacts including habitat modification and destruction, land contamination, waste production, water consumption and water pollution. At the global scale, concerns about the environmental impacts of aviation now focus sharply on the issue of climate change. The impact of air transport on global climate is already significant and is growing (Lee 2004; Mayor and Tol 2010). Aircraft emit the important greenhouse gas, carbon dioxide (CO_2), thereby contributing to the radiative forcing of climate. Besides the direct effects of CO_2 emissions, however, aircraft have various indirect effects on climate. Aircraft emissions of NO_x in the lower stratosphere and upper troposphere – the levels at which civil aircraft generally cruise – are responsible for the catalytic production of ozone (O_3) which acts as a powerful greenhouse gas at those levels, but NO_x emissions also deplete another greenhouse gas, methane (CH_4). Aircraft also emit soot and sulphate particles, which have different radiative effects on climate. In addition, aircraft create contrails (condensation trails) and cirrus clouds, the climate effects of which are not yet fully understood (Lee 2004). Further concerns have been expressed about the potential environmental impacts of a growing fleet of business jets, and even of a proposed fleet of supersonic aircraft cruising at high altitudes (in the mid-stratosphere), which could contribute to the destruction of stratospheric O_3 as a consequence of their NO_x emissions. Hence air transport has many environmental effects. Those effects interact in complex ways; they are rapidly increasing; and they can no longer be ignored by policymakers.

The environmental impacts of air transport form the subject of this book; they are discussed in greater detail in subsequent chapters. Here, it is simply acknowledged that those environmental impacts are compounded by the sustained, rapid growth of the air transport industry. Whilst technological and operational improvements have been made (such as advances in fuel combustor technology, the introduction of low-sulphur content fuels, the development of more aerodynamically efficient airframes and the use of noise abatement procedures), the growth of the industry has outpaced those improvements (Åkerman 2005; IPCC 1999; RCEP 2002; Sustainable Aviation 2005). Consequently, although greater fuel efficiencies – and concomitant emissions reductions – have been achieved, the absolute environmental impacts of air transport are increasing (Mayor and Tol 2010). Even in scenarios containing optimistic assumptions about the rate of technological progress, air transport is projected to produce almost twice as much CO_2 in 2030 as in 2002; in some scenarios, CO_2 emissions from aviation are projected to more than treble over the same period (Horton 2006; Lee 2004). By 2050, based on existing (Kyoto Protocol) commitments to reduce greenhouse gas emissions, aviation CO_2 emissions could consume entire national carbon budgets unless mitigation measures are urgently taken (Bows et al. 2006; 2009). In addition, whilst concerns about climate change represent an important obstacle to the growth of air transport, local environmental constraints may also be acute; at some airports, infrastructure development is already precluded by air quality legislation and by local agreements to limit aircraft noise levels. Furthermore, environmental regulations are likely to become more stringent as public tolerance of environmental impacts decreases. For all of these reasons, the air transport industry faces severe constraints to its

Table 1.1   The main environmental impacts of air transport

Environmental Impact	Main Causes
Aircraft noise	Aircraft operations Aircraft maintenance and engine testing Airport access traffic Airport stationary plant Airport surface vehicles
Air pollution	Aircraft emissions Emissions from airport access traffic Emissions from airport stationary plant Emissions from airport surface vehicles
Climate change	Aircraft emissions Contrails Aircraft-induced cirrus clouds Emissions from airport access traffic Emissions from airport stationary plant Emissions from airport surface vehicles Airport construction
Ecological change	Airport construction Coastal modification Drainage modification Watercourse modification
Habitat degradation	Airport construction Coastal modification Drainage modification Watercourse modification
Land contamination	Airport construction Airport waste disposal Aircraft servicing and maintenance Fuel, oil and hydraulic fluid spillage De-icing fluid run-off
Waste generation	Aircraft operations Airport operations
Water consumption	Aircraft operations Aircraft servicing and maintenance Airport operations
Water pollution	Aircraft servicing and maintenance Airport construction Airport waste disposal Fuel, oil and hydraulic fluid spillage De-icing fluid run-off

growth due to limited environmental capacity (Graham and Guyer 1999; Upham 2001; Upham et al. 2003; 2004).

Thus the air transport industry faces a dilemma: how to deliver vital economic and social benefits – benefits that are increasingly being demanded by consumers – whilst reducing its absolute environmental impacts. That dilemma represents a formidable challenge because of the projected rapid growth of demand for air travel and tourism, the strong links between air transport service provision and economic growth, the high abatement costs of the sector and the limited potential for radical technological solutions to be found over decadal timescales (DfT 2004; IPCC 1999). Some improvements in environmental performance are expected to be achieved by 2030 due to innovations in engine and airframe technologies, in aviation fuels and in operating systems and procedures (including more efficient air traffic management (ATM)). However, even cumulatively, those improvements are unlikely to offset the escalating impacts of a growing air transport sector. In the long term, radical technological solutions are required. Until such solutions are available, attempts to make air transport compatible with sustainable development must rely on the formulation and implementation of effective policy. Yet progress in developing policy for international air transport has been slow, partly because civil aviation is a highly competitive, cross-border activity that has been historically been regulated by myriad bilateral air service agreements (ASAs) between states as well as through broader international agreements (Pastowski 2003). The task of devising policy measures to reduce the environmental impacts of air transport involves negotiating sensitive issues of equity – both intragenerational and intergenerational – as well as balancing a wide range of economic, social and other environmental considerations. Such issues are now framed within the context of increasingly urgent calls to reduce the impacts of climate change. At the same time, the extent and sophistication of the scientific and technological knowledge required to inform critical decisions about climate change – and about other global issues – are becoming apparent. In all, the challenge facing policymakers is immense.

Responding to the Challenge

A central observation that informs this book is that, whilst air transport has a range of significant (and growing) environmental impacts, environmentalist concerns, in turn, are increasingly curtailing the growth of the air transport industry. Indeed, the expansion of some major airports is already tightly constrained for environmental reasons (Upham 2001; Upham et al. 2003; 2004). Therefore, the task of decoupling the growth of air transport from its environmental impacts is vital for both economic development and environmental protection. Such a decoupling relies both on the achievement of major technological and operational improvements in the environmental performance of aircraft and on the introduction of effective policy measures to reduce aviation environmental impacts. Technological responses include improvements in engine and airframe design and performance as well as the development of alternative fuels. However, given the high abatement costs that characterise the air transport sector, the long lead-in times associated with developing aviation infrastructure and the long in-service lifetimes of aircraft, there is, apparently, limited potential for radical technological progress to be made in the short to medium term – although those options may be more promising in the long term (DfT 2004; IPCC 1999). Operational responses involve different methods of loading, manoeuvring and maintaining aircraft in addition to the use of revised ATM systems and procedures. Operational measures, likewise, require substantial reforms – such as major revisions of airspace and of ATM systems and procedures – to be made if they are to drive substantial improvements in environmental performance, although modest improvements in efficiency may be achieved in the short to medium term. Policy responses, as in other areas of environmental governance, focus on the use of a wide range of policy instruments to align economic growth with the principles of sustainable development (Baker 2006). Such policy instruments include regulatory measures (standards), market-based measures (such as taxes, emissions charges, subsidies and tradable permits) and voluntary measures (such as carbon offsetting schemes). In principle, policy instruments can be devised and implemented relatively quickly, although progress in this respect has been slow to date (Pastowski 2003). In subsequent chapters of this book, the challenge of ensuring that aviation growth is compatible with sustainable development is considered in more detail, in terms of technological, operational and policy responses. A brief summary of those types of response is provided below.

In general, technological responses to the challenge of reducing the environmental impacts of air transport have focused on achieving improvements in engine and airframe design and performance, and on developing alternative aviation fuels (Thomas and Raper 2000). Efforts to achieve the former have centred on maximising the fuel efficiency of aircraft, both by reducing the weight and drag of airframes and by maximising the energy conversion efficiency of engines. Over a period of four decades, from around 1960, specific aircraft fuel efficiency (which is defined as fuel efficiency per passenger-kilometre) improved by around 70 per cent due to improvements in engine technologies and airframe design, and because of increased load factors (IPCC 1999). Further improvements in airframe performance are anticipated with ongoing improvements in aerodynamic efficiency, the use of advanced materials, innovation in control and handling systems and the development of radical aircraft designs (such as the blended-wing body). However, aircraft engine technologies are now relatively mature and the prospect of radical improvements in engine performance appears to be remote. Currently, the most fuel-efficient aircraft engines are high-bypass, high-pressure-ratio gas turbine engines, and the prospects for developing viable alternatives remain elusive (Lee 2004). One important consideration in this respect is that aircraft engine design may be optimised for fuel efficiency – and hence, as a side-effect, for minimising CO_2 emissions – or alternatively for reducing NO_x emissions, but not for both, meaning that a trade-off exists in the management of those emissions (IPCC 1999). Technological improvements in aviation fuels have resulted in the development of low sulphur fuels, leading to reduced emissions of sulphur oxides (SO_x), and recent research has investigated the potential for biofuels or hydrogen to supplement or replace kerosene. Overall, however, technological responses – both those relating to aircraft design and performance and to aviation fuels – require substantial investment and are likely to yield benefits only in the long term.

Operational responses are based on the principle of maximising fuel efficiency by a variety of means: reducing aircraft weight, increasing load factors, ensuring high levels of aircraft maintenance, minimising route distances, optimising cruising speeds and levels, and manoeuvring aircraft more efficiently (IPCC 1999). Loading aircraft efficiently involves a combination of (a) minimising the weight of the aircraft before its payload is stowed (for instance, by minimising the unusable fuel carried) and (b) maximising the payload. More frequent aircraft maintenance (especially engine maintenance) may help to ensure that fuel efficiency is maximised throughout the service lives of the airframe and engines. Environmental impacts may also be reduced through the use of revised ATM procedures: reduced cruising speeds; flight level optimisation for emissions reduction; arrival management (AMAN) and departure management (DMAN) systems; continuous descent approaches (CDAs) and 'low-power, low-drag' (LP/LD) approaches, which may reduce aircraft emissions and noise during descent; noise abatement departure procedures (NADPs) and noise preferential routes (NPRs), which are intended to reduce noise levels in the vicinity of airports; and expedited climb departure procedures, which could allow aircraft to climb rapidly to their optimal cruising levels (Dobbie and Eran-Tasker 2001; ICAO 2004). Environmental impacts may also be reduced by the use of hub-bypass route planning and by the use of fixed electrical ground power (FEGP) in preference to auxiliary power units (APUs) (Morrell and Lu 2006). However, despite the various ways in which operational procedures may be revised, their potential to reduce the environmental impacts of aircraft is relatively modest and 'these kinds of operational measures will not offset the impact of the forecast growth in air travel' (DTI 1996, 10).

Therefore, due to the high abatement costs of the aviation sector, and because of the limited potential for radical technological or operational solutions to be found in the short to medium term, success in meeting the challenge of reconciling aviation growth with environmental protection depends on the formulation and implementation of effective policy. Numerous instruments are available to policymakers, including regulatory, market-based and voluntary instruments; many of those have received scrutiny from a wide range of stakeholders. Thus proposals to cap aviation emissions, to impose taxes and emissions charges, to introduce or remove subsidies, to issue tradable permits for aviation emissions and to encourage the use of voluntary agreements, have been widely debated and contested (Bishop and Grayling 2003; IPCC 1999; Pastowski 2003). Such proposals have individual strengths but they are also problematic for reasons that are explained in more detail in subsequent chapters of this book. In general, however, regulatory approaches face the problem that air transport is an international industry that spans national jurisdictions, and that nations vary in their capacity to monitor and enforce environmental standards. Market-based approaches must negotiate complex issues relating to the varying competitiveness of air transport service providers – including the need to internalise varying costs of pollution – whilst facilitating access to international air transport markets on an equitable basis between nations. Voluntary approaches – such as the use of carbon offsetting schemes and the introduction of codes of conduct – face the criticism that they are too weak to catalyse the profound behavioural change that is required to align air transport with the principles of sustainable development.

The varying effectiveness, complexity and political acceptability of the different policy instruments means that no single instrument appears to be ideal, and the use of a combination of regulatory, market-based and voluntary approaches will probably be required in future aviation environmental policy (Pastowski 2003). Furthermore, policy options vary in their applicability at different geographical scales. Whilst local responses are required for the management of local environmental issues (air quality degradation, noise, habitat modification and destruction, water use and pollution, land contamination and waste generation) in the vicinity of airports, the global issue of climate change is likely to continue to dominate debates about aviation environmental impacts and about the growth of the industry, and that issue requires coordinated action at national and international levels (Stern 2007). In turn, policy approaches designed to deal with issues at national and international levels need to be aligned with many other policies. In the case of climate policy, national commitments to achieve reductions in greenhouse gas emissions could have profound implications for entire economies and societies: meeting such commitments may force fundamental changes to be made in the distribution and use of energy; in the development and availability of fuels; in infrastructure, business models, technologies and operating practices; and in the ways in which services are delivered. Meeting national emission reduction targets will certainly require a much greater degree of policy integration than currently exists. Aviation and climate policies should ideally be integrated with each other, and also with other policy frameworks – particularly sustainable development, energy, transport and other environmental policies – yet major disparities and contradictions currently exist between many of these policy areas (Bishop and Grayling 2003; Bows et al. 2006; 2009).

Thus the challenge faced by policymakers in attempting to reconcile the growth of air transport with the need for environmental protection should be placed within the context of a greater challenge: that of promoting sustainable development. Profound changes in economies, societies and environments worldwide are expected to occur as the extent and severity of human impacts on the global environment become apparent, and as human societies attempt to respond adequately to those impacts. The air transport industry, in common with other sectors, must increasingly adapt its activities to that changing context. The first steps have been tentative ones. To date, policy approaches have focused on the inclusion of international aviation within emissions trading schemes, such as the EU Emissions Trading Scheme (ETS), and on encouraging the use of voluntary agreements within the industry – typically, based on the use of carbon offsetting schemes and on commitments to achieve 'carbon neutrality'. However, if the environmental impacts of air transport are not sufficiently mitigated by those measures – and if environmentalist concerns continue to deepen – then policymakers will face intense pressure to curb the growth of the air transport industry. Subsequent measures to limit the environmental impacts of aviation could mean the imposition of more stringent emissions limits; the removal of existing privileges and subsidies of the industry; the wider use of emissions charges, fuel taxes and other levies; and, ultimately, the use of severe demand restraint measures (Stern 2007). Such measures would be extremely unpopular and could have many undesirable economic and social consequences. Therefore, the development of the air transport industry depends critically on the use of effective policy instruments to reduce aviation environmental impacts – and, ultimately, on finding technological solutions to mitigate the effects of aircraft emissions on climate.

Evolving Knowledge of Aviation Environmental Issues

Concerns about the environmental impacts of air transport are not new; they have been documented for more than four decades (Lee and Raper 2003). Analysis of the literature of aviation environmental issues suggests that those issues have attracted increasing attention alongside the emergence of environmental concerns – and of environmentalism – more generally (Pepper 1996). Aviation environmental impacts were not sufficiently prominent to merit discussion in Sealy's (1966) *Geography of Air Transport*. However, the problem of aircraft noise had prompted a certain amount of public concern since the 1950s (Freer 1994). In 1966, the environmental problems caused by civil aircraft were discussed at a conference in London involving representatives of twenty-six nations and eleven international aeronautical organisations; the main aviation environmental issue at that time was considered to be aircraft noise (Price and Probert 1995; Stratford 1974). Several studies of the impacts and management of aircraft noise were published in a 1967 edition of the *Journal of Sound and Vibration* (Hubbard et al. 1967; Kryter 1967; Lauber 1967; Pattarini 1967; Sawyer 1967). In 1969, the International Civil Aviation Organization (ICAO) established its Committee on Aircraft Noise (CAN), which developed international standards for aircraft noise performance and which focused on encouraging technological developments leading to the manufacture of quieter engines, such as the use of acoustic linings (Mangiarotty 1971). Thus one environmental impact of air transport has been documented – and has been regulated, at least at the point of manufacture – for around forty years.

Besides aircraft noise, other aviation environmental issues were identified at a relatively early stage, including the pollution arising from aircraft engine exhaust emissions, de-greasing agents, runway and aircraft de-icing compounds and fire extinguishing substances (Price and Probert 1995). In particular, the dark exhaust plumes of the early turbojet-powered aircraft were easily visible and generated concerns about local air pollution. In 1971, ICAO adopted Annex 16 (Environmental Protection) to the 1944 Chicago Convention, which covered both emissions and noise (Freer 1994, 30). Shortly afterwards, in 1972, at the United Nations Conference on the Human Environment (UNCHE) in Stockholm, ICAO adopted a resolution to investigate the impacts of aviation on the quality of the human environment – a course of action which led to the publication, in 1977, of the ICAO Circular 134-AN/94, *Control of Aircraft-Engine Emissions*. In that document, the ICAO proposed methods to control vented fuel, smoke and other pollutants for new subsonic aircraft engine designs (Mortimer 1979; Price and Probert 1995). Concerns about localised air pollution also prompted the US Environmental Protection Agency (EPA) to regulate aircraft emissions; whilst that early legislation applied only in the US, it was the forerunner of the international ICAO engine certification standards that remain in use today (ICAO 1993; 2005; IPCC 1999; Lee and Raper 2003). In 1977, ICAO established the Committee on Aircraft Engine Emissions (CAEE) and, by 1981, had established standards for three pollutants emitted by aircraft: carbon monoxide (CO), unburned hydrocarbons (HCs) and NO_x. Limits were also placed on emissions of smoke, and the intentional venting of fuel from engines was prohibited.

Although early legislation focused on localised aviation environmental impacts (localised air pollution and noise), the effects of aircraft on the global atmosphere were also considered during the 1970s (RCEP 1971; 2002). Initially, those effects were mainly hypothetical: they related to a proposed fleet of supersonic aircraft cruising in the stratosphere. Studies by Crutzen (1971) and Johnston (1971) indicated that the NO_x emissions from such a fleet could significantly deplete stratospheric O_3; however, only a limited supersonic fleet of supersonic civil aircraft was eventually developed (comprising Concorde and the Tupolev Tu-144) and the expected large supersonic fleet was never built (Lee and Raper 2003; Rogers et al. 2002). Nevertheless, by 1971, the literature of aviation environmental impacts covered the now-familiar three main issues: aircraft noise, localised air pollution and impacts on the global atmosphere (although the specific issue of climate change had not yet become prominent). It was not long before the possibility that environmental issues could constrain airport capacity received scrutiny: Ferrar (1974) investigated the impacts of noise and air pollution on airport capacity and derived some implications for airport management. Related issues of fuel optimisation – including a range of operational measures to increase fuel efficiency (such as reducing the practice of fuel tankering) – and the influence of aircraft size on operating costs (including effects on fuel consumption) were also investigated (Drake 1974; Nicol 1977).

In 1983, ICAO established the Committee on Aviation Environmental Protection (CAEP), which superseded the CAN and the CAEE; the work of CAEP continues today and includes most of the environmental activities now undertaken by ICAO. Aircraft noise remained an issue of concern during the 1980s and 1990s, with an increasing emphasis on the subjective responses of individuals to aircraft noise (Job 1996; Ko and Lei 1982). Interest in aviation environmental impacts was substantially renewed in the late 1980s and early 1990s, however, with the emergence of a new environmental issue: climate change. In particular, some scientists realised that the cruising levels of subsonic aircraft (in the upper troposphere and lower stratosphere) are critical for global climate, both in terms of atmospheric chemistry and climate sensitivity. Concerns about the impacts of aircraft on global climate focused initially on the role of NO_x in generating O_3 in the upper troposphere and the lower stratosphere, since O_3 at those levels acts as a powerful greenhouse gas (Rogers et al. 2002). Since then, various climate impacts of aircraft have been investigated, including the formation of contrails and cirrus clouds as well as certain other effects of particle emissions (IPCC 1999; Lee and Raper 2003; Lee 2004; Schumann 1996; Schumann and Wending 1990).

In 1992, ICAO compiled an inventory of the environmental problems associated with civil aviation (Table 1.2). Those problems included aircraft noise, localised air pollution, global atmospheric effects, impacts of airport and infrastructure construction, water and soil pollution, waste generation, and environmental degradation resulting from aircraft accidents and incidents (Crayston 1992; ICAO 2001; Price and Probert 1995). By that time, it had become apparent that the air transport industry faced a growing challenge in responding adequately to environmental concerns (Price and Probert 1995; RCEP 1994; 2002). Andrieu (1993) argued that airlines would have to meet increasingly tough environmental standards and Somerville (1993, 173) acknowledged that, whilst environmental factors were already significant for the industry, they were likely to become 'dominant' and 'of overriding importance' in the future. Numerous overviews, assessments and syntheses of aviation environmental issues were produced around that time (Brasseur et al. 1998; Friedl et al. 1997; Price and Probert 1995; Schumann 1994; Wahner et al. 1995). In stark contrast to Sealy's (1966) earlier work on the subject, Graham (1995) devoted an entire chapter of his text, *Geography and Air Transport*, to the subject of air transport and the environment. In addition, during the 1990s, greater efforts were made by the air transport industry to publicise its environmental initiatives (Chaplin 1996; IATA 1996; Ralph and Newton 1996), whilst an increasing number of reports and studies of specific aviation environmental issues – as well as of proposed policy responses – were published (CEC 1999; Crayston and Hupe 1999; Dobbie 1999, FAA 1997; Gander and Helme 1999; Graham and Guyer 1999; ICAO 1999; Janić 1999; Mato and Mufuruki 1999; Norgia 1999; Nygardis 1999; Simpson and Kent 1999; Vedantham and Oppenheimer 1998; Vincendon and von Wrede 1999; Walle 1999).

Table 1.2   ICAO inventory of aviation environmental problems

Environmental Impact	Examples
Aircraft noise	Aircraft operations Engine testing Airport sources Sonic boom (due to supersonic aircraft)
Local air pollution	Aircraft engine emissions Emissions from airport motor vehicles Emissions from airport access traffic Emissions from other airport sources
Global phenomena	Long-range air pollution (e.g. acid rain) The greenhouse effect Stratospheric ozone depletion
Airport/infrastructure construction	Loss of land Soil erosion Impacts on water tables, river courses and field drainage Impacts on flora and fauna
Water/soil pollution	Pollution due to contaminated run-off from airports Pollution due to leakage from storage tanks
Waste generation	Airport waste Waste generated in-flight Toxic materials from aircraft servicing and maintenance
Aircraft accidents/incidents	Accidents/incidents involving dangerous cargo Other environmental problems due to aircraft accidents Emergency procedures involving fuel dumping

Source: Adapted from Crayston (1992, 5); Price and Probert (1995, 140)

Efforts to document the atmospheric impacts of aircraft culminated in the publication, in 1999, of a Special Report of the Intergovernmental Panel on Climate Change (IPCC), *Aviation and the Global Atmosphere* (IPCC 1999). This was a landmark document: it was the first sectoral account produced by the IPCC and it contained estimates of the radiative forcing of climate due to various aircraft emissions (Lee and Raper 2003, 78; Rogers et al. 2002). The IPCC (1999) report provided a detailed assessment of the nature and severity of the atmospheric impacts of aircraft – and of the magnitude of projected future impacts – and it established an influential framework for further research into aviation environmental issues. The IPCC (1999) report has defined and shaped the research landscape and has influenced the development of aviation environmental policy. Since the publication of that authoritative report, other important documents have been produced. Rogers et al. (2002) provided an updated account of aviation impacts on local air quality and on global climate, identifying areas of scientific uncertainty and emphasising the need for further research. Subsequently, various aspects of the relationship between air transport and sustainable development were explored in a multidisciplinary book, *Towards Sustainable Aviation*, which incorporated a range of scientific, technical, engineering, policy and environmental management perspectives towards aviation environment issues (Upham et al. 2003). In that book, a framework was presented for considering the economic and social benefits of air transport alongside its environmental and social costs. Those notable overviews have been supplemented with a large number of scientific, economic and policy-focused studies of specific aspects of aviation environmental issues, many of which inform the subsequent chapters of this book. Knowledge of aviation environmental issues continues to evolve rapidly, particularly in relation to climate change, as researchers and other commentators explore the various options for mitigating the effects of aviation emissions on climate.

Aims, Approach and Outline of this Book

The subject of air transport and the environment has already received considerable attention from many commentators, but a new account is timely due to the rapid pace of developments both in the air transport industry and in environmental science. In particular, understanding of the scientific, economic, social and political aspects of climate change has dramatically improved and new concepts and practices have emerged, such as carbon management, carbon footprints, carbon neutrality and carbon offsetting. New insights into the environmental impacts of air transport, their relative importance, the trade-offs between those impacts and the implications for the growth of the industry are also available. Given those recent developments, the aims of this book are: (a) to provide an updated overview of the environmental issues associated with air transport, emphasising recent scientific and policy developments; (b) to derive implications for the operation, growth, development and management of the air transport industry; (c) to inform aviation environmental policy; and (d) to provide a new analysis of the relations between air transport, environmental protection and sustainable development, highlighting some areas in which further research is required. Overall, this book attempts to provide a constructive, policy-relevant synthesis of a wide range of perspectives rather than advocating one particular viewpoint.

This account focuses on presenting the relevant environmental issues rather than on describing technological or operational systems or procedures in detail (although the use of some technical and operational terms is unavoidable). Frequent use is made of some terms that abound in environmental (and other) literature, including 'sustainable development' and 'environmental degradation'; it is worth emphasising that such terms are highly complex and contested (Adams 2009; Baker 2006; Dresner 2008; Elliott 2006; Redclift 1987; WCED 1987). Given the historical evolution of air transport in the Northern Hemisphere, centred on the air transport markets of North America and Europe, those regions receive most attention in this book; other regions – such as South America and Africa – receive scant attention in comparison. However, the current and projected growth of air transport in other parts of the world – especially in China, Southeast Asia and parts of the Middle East – is one of the most significant trends in the contemporary geography of air transport. If current trends continue, that growth is likely to be associated with severe environmental impacts; thus the growth of demand for air travel in Asia could have profound implications for the management of aviation environmental impacts (Mayor and Tol 2010).

The book begins by covering some contextual material in Chapter 2, including an explanation of the nature of aviation emissions and of the monitoring and modelling techniques used by scientists to understand aviation emissions and their impacts. The growth of air transport is also considered in Chapter 2, for that growth represents a crucial dimension of the overall environmental impact of aviation. Subsequent chapters consider the three most important aviation environmental issues: impacts on climate (Chapter 3), impacts on air quality (Chapter 4) and aircraft noise (Chapter 5). In those chapters, the nature of each respective environmental impact is discussed, followed by an account of the various options for reducing that impact. The environmental impacts of air transport are then considered in a broader perspective – in relation to sustainable development – in Chapter 6; from that perspective, the need to balance the economic, social and environmental costs and benefits of air transport is considered, together with some ways in which the industry may be aligned more closely with the principles of sustainable development. Finally, Chapter 7 contains a summary of the main points emerging from the preceding chapters.

2 Understanding the Issues

Introduction

How do aircraft interact with their environment? What interactions occur between airports and the environment? How do those interactions vary geographically and with time? Answering such questions is central to understanding aviation environmental impacts. In particular, two aspects of the relationship between air transport and the environment are important in understanding current debates: (a) the emissions from aircraft and airport infrastructure; and (b) the growth in demand for air transport. Those two themes – aviation emissions and aviation growth – recur throughout the subsequent chapters of this book, and understanding the issues associated with each is vital for a balanced view of the environmental impacts of air transport. Explaining the main trends in aviation emissions and growth is the focus of this chapter. Aircraft and airports have many interactions with the environment, by far the most important of which are the emissions generated by aircraft during their operation. This is not to imply that other environmental interactions are negligible. The materials and energy used in manufacturing, maintaining and ultimately disposing of aviation infrastructure – over the life cycle of that infrastructure – may be substantial and should be considered in aviation environmental impact assessments, especially in the case of airport infrastructure development. In addition, many environmental interactions occur during the surface transport of employees, passengers, freight, goods and waste to, from and within airport sites, and during the servicing of aircraft before and after flights. At the local scale, such interactions may result in significant, cumulative environmental impacts because major airports are sites of intensive resource consumption and waste production. Nevertheless, at the global scale – and increasingly at the scale of individual airports – the impact of aircraft emissions far outweighs any other aviation environmental effect in its magnitude and significance (Bows et al. 2009; Upham 2003). Therefore, an understanding of aircraft emissions is crucial to any discussion of aviation environmental impacts. Furthermore, despite some impressive technological and operational advances in the efficiency of air transport operations, absolute aircraft emissions are increasing – and are projected to continue to rise – due to the growth in demand for air transport. This makes an understanding of the trends in aviation growth critical for a full understanding of aviation environmental impacts.

This chapter focuses on aviation emissions – predominantly those due to aircraft – and on aviation growth in turn. It begins with an overview of the kerosene combustion process and describes the most important chemical species emitted in the exhaust of aircraft engines. The full significance of those pollutants is explained in subsequent chapters, where their impacts on climate and on air quality are discussed in more detail. However, some fundamental concepts – such as the emission factor of a pollutant – are introduced in this chapter. Some of the techniques by which scientists investigate aircraft emissions are also introduced; those techniques include the monitoring and modelling of aircraft emissions and the production of global aircraft emission inventories. Next, this chapter turns to the subject of aviation growth, especially the projected growth in demand for air transport. The main trends in aviation growth are outlined, based on several authoritative analyses and on some influential market forecasts, and the important distinction between forecasts and scenarios is explained. This account highlights the links between the growth of air transport and three interrelated trends: economic growth, globalisation and the growth of tourism. For many years, sustained economic growth, globalisation and tourism growth meant that the future of air transport seemed likely to be characterised by rapid, sustained expansion. However, the recent emergence of climate change as a dominant global environmental issue has altered the terms of the debate. If it were unconstrained, the growth of air transport would transform the industry from being (currently) a modest contributor to climate change to being (by 2050) a very substantial polluter (Bows et al. 2009; Gössling and Upham 2009; RCEP 2002). Given the long lead-in and in-service times associated with aviation infrastructure, the need to address those projected impacts has already become urgent. Consequently, as this and subsequent chapters explain, the task of managing aviation environmental impacts is predominantly concerned with the management of aviation emissions and of aviation growth.

Understanding Aviation Emissions

Modern commercial transport aircraft are powered by the combustion of kerosene in turbofan and turboprop gas turbine engines. Of those two engine types, turbofans dominate the global fleet (Lee 2004; RCEP 2002). Kerosene is a combustible hydrocarbon mixture obtained from the fractional distillation of petroleum; it contains a large variety of carbon chain molecules, generally with chain lengths of nine to 16 carbon atoms in the case of Jet A-1 fuel (RCEP 2002). For the purposes of civil aviation, kerosene is produced to internationally standardised specifications and is combined with various additives to create conventional jet fuels. During the operation of an aircraft engine, fuel is injected into the combustor where it is mixed with compressed air and ignited. Thus combustion occurs under conditions of increased temperature and pressure (due to the compression of the air) and the burning of the fuel further raises the temperature of the pressurised mixture. The hot, compressed gas produced by the combustor is then used in the engine turbine section to drive the compressor (thereby reducing the temperature and pressure of the gas) and to propel the aircraft. The majority of modern commercial transport aircraft are equipped with high bypass, high pressure ratio turbofan engines in which a second turbine is used to create a bypass jet that generates the propulsive force. In order to fly efficiently, aircraft require engines with a high power output to weight ratio, a requirement that imposes limitations on the designs that can be used in aviation compared with, for instance, land-based power generation (IPCC 1999). After more than six decades of development, aircraft gas turbine technology is now relatively mature and the prospects of radical new designs emerging in the short term are remote (RCEP 2002).

In ideal conditions, kerosene undergoes complete combustion to produce carbon dioxide (CO_2) and water vapour (H_2O) in proportions that depend upon the specific carbon to hydrogen ratio of the fuel. At the same time, during complete combustion, a very small proportion of sulphur dioxide (SO_2) is produced as a result of the oxidation of sulphur-containing compounds that are added to the fuel to improve its lubricity. In addition to those combustion products, large quantities of ambient air – mainly nitrogen (N_2) and oxygen (O_2) – pass through the engine (Figure 2.1). Hence, when aircraft are cruising, their combustion products constitute only around 8.5 per cent of the total mass flow leaving the engine (Figure 2.2). In addition to the species mentioned above, many other substances are emitted as a consequence of the incomplete (non-ideal) combustion of the fuel; those residual species include nitrogen oxides (NO_x), hydrocarbons (HCs), carbon monoxide (CO), soot particles and sulphur oxides (SO_x). The last of those species, SO_x, gives rise to sulphate particles (IPCC 1999; RCEP 2002). Overall, however, in order to achieve their high power output to weight ratio, aircraft engines are by necessity extremely efficient in converting the chemical energy of the fuel to kinetic energy (Rogers et al. 2002). Thus, when aircraft are cruising, only around 0.4 per cent by volume of the combustion products contains the residual products of non-ideal combustion (Figure 2.2). The large majority of that residue comprises NO_x (IPCC 1999). Nevertheless, although the residual products of incomplete combustion make up only a small proportion of total aviation emissions, their environmental impacts may be disproportionately large, as subsequent chapters will show. Furthermore, the quantities and proportions of some aircraft emissions vary widely during the different phases of flight under different engine operating regimes (Table 2.1).

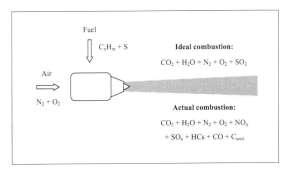

Figure 2.1   Aircraft   engine   emissions   under   ideal   and   actual conditions

Note: C_nH_m, kerosene; CO, carbon monoxide; CO_2, carbon dioxide; C_{soot}, soot particles; H_2O, water vapour; HCs, hydrocarbons; N_2, nitrogen; NO_x, nitrogen oxides; O_2, oxygen; S, sulphur; SO_2, sulphur dioxide; SO_x, sulphur oxides

Source: Adapted from IPCC (1999, 235)

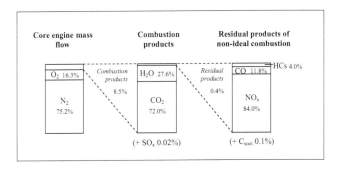

Figure 2.2   Kerosene combustion products under actual conditions

Note: CO, carbon monoxide; CO_2, carbon dioxide; C_{soot}, soot particles; H_2O, water vapour; HCs, hydrocarbons; N_2, nitrogen; NO_x, nitrogen oxides; O_2, oxygen; SO_x, sulphur oxides

Source: Adapted from IPCC (1999, 235)

Table 2.1   Aircraft emissions under different engine operating regimes

Species	Emissions (g) per kg kerosene burned		
	Idle	**Take-off**	**Cruise**
CO_2	3,160	3,160	3,160
H_2O	1,230	1,230	1,230
NO_x (as NO_2)			
Short-haul	4.5 (3–6)	32 (20–66)	7.9–11.9
Long-haul	4.5 (3–6)	27 (10–53)	11.1–15.4
CO	25 (10–65)	<1	1–3.5
HC (as methane)	4 (0–12)	<0.5	0.2–1.3
SO_x (as SO_2)	1.0	1.0	1.0

Source: Adapted from IPCC (1999, 235)

To simplify the analysis, the components of the ambient air that pass unaltered through the engine may be disregarded. During flight, therefore, aircraft engines emit gases (CO_2, H_2O, NO_x, HCs, CO and SO_x) and particles (mainly soot and sulphate particles) directly into the atmosphere (Figures 2.1 and 2.2). Those emissions alter the chemical composition of the atmosphere both directly and indirectly, giving rise to a variety of environmental impacts (RCEP 2002). The unique property of aircraft emissions is that, during cruise, they are injected into the upper troposphere and lower stratosphere, around nine to 13 kilometres above the surface of the Earth, where they form the predominant anthropogenic emissions (RCEP 2002). As the IPCC (1999) has acknowledged, the effects of most aircraft emissions depend strongly on the flight altitude and on whether aircraft fly in the troposphere or stratosphere. The atmospheric effects of aircraft emissions may be markedly different at typical cruising levels from the effects of the same substances at ground level. At cruising levels, aviation emissions have several significant effects: they change the concentration of atmospheric greenhouse gases, including CO_2, ozone (O_3) and methane (CH_4); they can cause condensation trails (contrails) to form in certain conditions; and they can increase cirrus cloud coverage (IPCC 1999). Aircraft emissions also have several effects closer to the ground during the landing and take-off (LTO) cycle: they alter local concentrations of NO_x, O_3 and particles in addition to their more general effect of elevating CO_2 concentrations. The main species emitted by aviation are briefly discussed in turn overleaf.

Carbon Dioxide

Carbon dioxide (CO_2) is emitted by aircraft in direct proportion to the quantity of fuel burned, with 3,160g (± 60g) of CO_2 being released per kilogram of kerosene burned (IPCC 1999; Lee 2004; 2009; RCEP 2002; Figures 2.1 and 2.2; Table 2.1). CO_2 is a greenhouse gas that occurs naturally in the environment, but it is also the single most important waste product of industrialised economies. If considered purely in terms of its radiative effect, CO_2 is not the strongest greenhouse gas, but its relative abundance and its very long atmospheric lifetime mean that its role in the global climate system is of overriding importance and it is one of the key pollutants covered by the Kyoto Protocol (IPCC 2007; Seinfeld and Pandis 2006; Stern 2007; UN 1998). Since CO_2 is a conservative gas that persists in the atmosphere over long periods, it becomes well mixed and globally distributed, and aviation CO_2 emissions become indistinguishable from the same quantity of CO_2 emitted by any other source (Archer 2005; IPCC 1999; Lenton et al. 2006; RCEP 2002; Seinfeld and Pandis 2006). Global aviation CO_2 emissions in 2005 were estimated to be 733Tg, which constituted roughly 2 to 2.5 per cent of total anthropogenic CO_2 emissions (IPCC 1999; Lee 2009; Peeters and Williams 2009). However, global aviation CO_2 emissions are increasing rapidly as a result of the sustained growth of air transport (Figure 2.3). By 2030, even taking account of technological and operational improvements, aviation CO_2 emissions are expected to have increased dramatically; by 2050, in some scenarios, those emissions are projected to have grown by a factor of three over 1992 levels (Eyers et al. 2004; Horton 2006; Lee 2004; Mayor and Tol 2010; Peeters and Williams 2009; RCEP 2002; Stern 2007; WRI 2005; Yamin and Depledge 2004; Figure 2.4). Given recent trends, by 2050, aviation CO_2 emissions could consume a large proportion of national carbon budgets – or even entire national allowances – depending on the particular scenarios considered (Bows et al. 2009; Owen and Lee 2006).

Water Vapour

Water vapour (H_2O), like CO_2, is emitted by aircraft in direct proportion to the quantity of kerosene burned, with around 1,230g (± 20g) of H_2O being released per kilogram of kerosene burned (IPCC 1999; Lee 2009; RCEP 2002; Figures 2.1 and 2.2; Table 2.1). Water vapour is a powerful greenhouse gas that plays a critical role in the natural greenhouse effect of the Earth (Seinfeld and Pandis 2006). The lower part of the atmosphere (the troposphere) is relatively humid as a result of natural hydrological

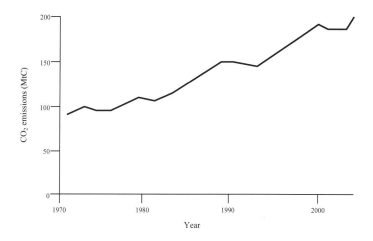

Figure 2.3   Global CO$_2$ emissions from aviation, 1970–2006

Source: Adapted from Bows et al. (2009, 12)

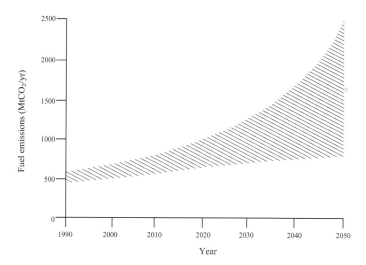

Figure 2.4   Global   CO$_2$ emissions   from   aviation   under   various scenarios, 1990–2050

Source: Adapted from Lee (2009, 31)

processes, and the quantity of water vapour emitted by subsonic aircraft is very small compared with ambient concentrations, even when the additional projected emissions for 2025 and 2050 are taken into account (Eyers et al. 2004; Lee 2009). Most aviation emissions of water vapour occur in the troposphere; once emitted, they are rapidly removed by precipitation over a timescale of weeks. A smaller proportion of water vapour is emitted in the lower stratosphere where it can accumulate, although its radiative effect is nonetheless smaller than that of other aircraft emissions such as CO_2 and NO_x (IPCC 1999). Therefore, the climate effects of water vapour emissions from civil aircraft are relatively small (Lee 2004; 2009; Lee and Raper 2003). However, the middle and upper parts of the stratosphere are extremely dry, meaning that any introduction of water to the stratosphere at higher levels is likely to increase atmospheric warming. Scenarios in which proposed fleets of supersonic aircraft operate at higher levels in the stratosphere could result in water vapour emissions from aircraft having a much larger effect on climate (Lee 2004; 2009; Lee and Raper 2003).

Nitrogen Oxides

Nitrogen oxides (NO_x) comprise nitric oxide (NO) and nitrogen dioxide (NO_2); they are created in aircraft engines as a result of several complex chemical pathways (Lee 2009). In general, however, NO_x is formed by the oxidation of atmospheric nitrogen in the high temperature conditions occurring in the combustor, with some small quantities of NO_x arising from the nitrogen content of the fuel (Rogers et al. 2002). As Lee (2004; 2009) has acknowledged, NO_x formation is not directly proportional to fuel burn but is instead a complex function of combustion temperature, pressure and combustor design (Figures 2.1 and 2.2; Table 2.1). Most of the NO_x emitted by aircraft is in the form of NO, but this is rapidly converted to NO_2 in the atmosphere; however, a small amount of NO_2 (termed 'primary NO_2') is emitted directly by the engines (Lee 2004; Lee and Raper 2003). The formation of NO_x during the combustion process is unavoidable, although NO_x emissions can in principle be reduced by using particular combustor designs (Lee 2009; Rogers et al. 2002). In fact, however, specific NO_x emissions from aircraft have tended to increase rather than decrease because efforts to improve the fuel efficiency of aircraft engines have prompted the use of greater engine pressure ratios, which in turn have led to higher temperatures and pressures in the combustors, and thus to increased NO_x production. Given the ongoing efforts made by engine manufacturers to improve fuel efficiency, aircraft NO_x emissions have been forecast to increase by a factor of approximately 1.6 over the period 2002–2025 (Eyers et al. 2004; Lee 2004). Yet concerns about the environmental impacts of NO_x emissions – especially on air quality in the vicinity of major airports – have prompted policymakers to call for simultaneous reductions in aircraft NO_x emissions and fuel efficiency improvements (ACARE 2004). The drive to limit NO_x emissions now focuses on the development of new combustor technologies, although the reduction of NO_x emissions may be achieved only at the expense of increased aviation CO_2 emissions, for a trade-off exists in the reduction of those two pollutants (Lee 2009). Besides their impacts on air quality, aircraft NO_x emissions have two significant, indirect effects on climate: the enhancement of tropospheric ozone (O_3) and the reduction of methane (CH_4), both of which are greenhouse gases. Those effects are discussed in more detail in Chapter 3.

Sulphur Oxides

Sulphur oxides (SO_x) are emitted by aircraft as a result of the oxidation of sulphur-containing compounds that are added to the fuel in low concentrations to improve its lubricity; thus aircraft SO_x emissions are directly proportional to the sulphur content of the fuel (IPCC 1999; Lee 2009). Most of the sulphur present in the fuel is thought to be emitted in the form of sulphur dioxide (SO_2), which is acknowledged to be an important local air pollutant. However, some oxidation through to S^{VI} species – such as sulphur trioxide (SO_3) and gaseous sulphuric acid (H_2SO_4) – occurs within the engine and the exhaust plume, and the emission of sulphur-containing species is believed to be important in the formation of volatile particles (IPCC 1999; Lee 2004; 2009). Overall, the quantities of SO_2 emitted by aircraft are very small in comparison with ambient concentrations, and the main importance of aircraft-derived SO_x is as a source of sulphate particles (Figures 2.1 and 2.2; Table 2.1). Sulphate particles have several important environmental effects including a role in the formation of contrails (see below).

Hydrocarbons

Hydrocarbons (HCs) are molecules consisting only of carbon and hydrogen atoms. Hydrocarbon emissions from aircraft include unburned molecules from the kerosene fuel mixture, sometimes termed 'unburned hydrocarbons' (UHCs), which occur in the exhaust plume when fuel molecules escape the flame zones and pass through the engine along with the rest of the mass flow (Figures 2.1 and 2.2; Table 2.1). However, a wide range of other HCs that form as intermediate species during chemical reactions within the engine may also be emitted by aircraft. Whilst some HCs are emitted during the operation of the engines, others are released into the atmosphere during refuelling between flights; those fugitive emissions are responsible for the characteristic odour that occurs in the vicinity of airports (DfT 2006; Winther et al. 2006).

Particles

Particles (sometimes termed 'particulate matter', PM) emitted from aircraft engines tend to be very small, with typical sizes in the aerodynamic diameter range 3nm to 4μm (Lee 2004, 5). They can be classified as either non-volatile or volatile. Non-volatile particles are mainly carbonaceous material (soot) formed in the primary combustion zone of the engine due to incomplete combustion of the fuel; most of the soot particles that form in the engine are then burned out in further combustion, although some small particles survive to be emitted in the exhaust (Figures 2.1 and 2.2). The average size of the soot particles emitted is very small: approximately 30–60nm (Lee 2004; 2009; Petzold et al. 1999). Volatile particles are mostly sulphate, although organic material also constitutes a smaller proportion of that group. Sulphate particles form in the exhaust plume (and possibly in the engine itself) as a result of the oxidation of sulphur-containing compounds added to the fuel; sulphate particles in the plume are roughly one to two orders of magnitude more abundant than soot particles (Lee 2004; Lee and Raper 2003). Particles therefore represent a heterogeneous group of emitted species that form both directly (during combustion) and as a result of chemical processes in the plume. Airborne sulphate particles and soot particles are both examples of aerosols: suspensions of fine solid or liquid particles in a gas (IPCC 1999; Seinfeld and Pandis 2006). Aerosols may be emitted directly as particles ('primary aerosol') or may form in the atmosphere by conversion processes ('secondary aerosol'). Soot and sulphate particles have small, direct impacts on climate by absorbing or scattering solar radiation (Seinfeld and Pandis 2006); those effects will be discussed in Chapter 3. In addition to those direct effects, particles also have indirect effects on climate through their role as cloud condensation nuclei in the formation of contrails and cirrus clouds. Whilst technological improvements have helped to reduce particle emissions from aircraft engines, such progress is now difficult to sustain and particle emissions are expected to increase over the period 2002–2025 (Eyers et al. 2004).

Other Emissions

In addition to the gases and particles described previously, many other substances are emitted by aircraft in small quantities (Anderson et al. 2006; Price and Probert 1995). Carbon monoxide (CO) is a gas formed during the incomplete combustion of fossil fuels, representing carbon that has not been fully oxidised to CO_2. However, the high efficiency of aircraft engines results in extremely low CO emissions during most aircraft operations (DfT 2006; Figures 2.1 and 2.2; Table 2.1). Other species are emitted only in trace quantities. Hydroxyl radicals (OH) are produced during the combustion process and participate in reactions involving the conversion of NO_x and sulphur-containing species to their oxidised forms. Although relatively few measurements of OH emissions have been made, OH is thought to have a concentration of around 1 ppb (part per billion) in the engine plume (Lee 2004). Nitrous acid (HONO) and nitric acid (HNO_3) emissions also occur, although the abundance of those species is estimated to be small (Lee 2004). Another group of trace species emitted by aircraft is termed 'chemi-ions' (CIs); these are gaseous ions formed at high temperatures by the chemiionisation of free radicals. CIs found in the exhaust plume include both negative and positive ions; they may be involved in the formation and growth of charged droplets in the atmosphere (Lee 2004; Rogers et al. 2002; Wohlfrom et al. 2000).

For many types of emitted species, an emission factor may be derived, which indicates the quantity of the gas or particle emitted as a result of the combustion of a standard quantity (1 kg) of kerosene (Peeters and Williams 2009). For some substances, the emission factor is constant regardless of how the engine is operated, whilst other emission factors vary depending on the power setting used (and on other variables). Thus CO_2 and H_2O emissions increase in direct proportion to fuel consumption and their emission factors are constants (see Table 2.1). In contrast, NO_x is produced more abundantly at high power settings for a fixed quantity of fuel burned, and the emission factor for NO_x also increases with a range of other variables in addition to the power setting, including the ambient conditions of temperature, pressure and humidity; the effects of the forward speed of the aircraft; and the extent of engine and airframe deterioration (Curran 2006; DfT 2006; Table 2.1). The emission factor of soot is also highest at high power settings. For HCs and CO emissions, their respective emission factors vary inversely with combustion efficiency (although not in a linear fashion) because they are the products of incomplete combustion. Hence aircraft emissions of HCs and CO are greatest at low power settings due to the fact that aircraft engines are optimised for maximum fuel efficiency at the relatively high power settings used during the cruise. SO_x emissions occur in direct proportion to the sulphur content of the fuel; therefore, for a given fuel type, the emission factor for SO_x is constant (Table 2.1). For almost all common aircraft engine types, emission factor values for regulated pollutants (CO, HCs, NO_x and smoke) and fuel flow data are published in the ICAO Engine Emissions Databank (CAA 2006). Significantly, however, those data are values measured at the point of engine certification, and actual engine performance during aircraft operations may vary considerably from those 'ideal' values (Curran 2006; DfT 2006; Eyers et al. 2004; Schürmann et al. 2007; Unique 2004). A recent study by Schürmann et al. (2007) has demonstrated that aircraft emission factors vary considerably with engine type – and they may also vary substantially from the emission factors published in the ICAO Engine Emissions Databank.

Whilst emissions from aircraft are the dominant type of aviation emissions, the operation of airports – and the transport of employees, passengers, freight, goods and waste to, from and within airport sites – also cause the emission of gases and particles to the atmosphere. In general, those emissions are typical of any fossil fuel combustion process. Therefore, airports and their ancillary vehicles potentially emit CO_2, H_2O, NO_x, SO_x, HCs, CO and particles, with the relative proportions of ideal and residual combustion products being dependent upon the efficiency of the processes used. However, unlike aircraft, most of which are designed to burn conventional kerosene according to tightly-controlled standards, airports have a certain degree of influence over the source of their energy supplies and over their on-site combustion activities. Thus surface transport vehicles may be fitted with catalytic converters to limit emissions of NO_x, HCs and CO, and airside support vehicles may be powered entirely by electricity or biofuels. Transport systems to, from and within airport sites may use electrical power or magnetic propulsion systems. Airport buildings and equipment may also be heated, air-conditioned and powered by 'green' electricity (electricity obtained from renewable sources) rather than by generators burning fossil fuels. Whilst some of these practices result in absolute emissions reductions, it is worth emphasising that others, such as the use of externally-generated electricity, do not necessarily reduce emissions but may simply transfer them to sites further from the airport.

Scientists use a variety of methods to understand aviation emissions. Direct monitoring of aviation emissions is in principle the most accurate way to investigate the abundance and properties of emitted species, but there are significant obstacles to direct monitoring: the extreme conditions existing in the aircraft engine and in the exhaust plume, which are hostile to sensitive scientific instruments; the fact that the engine and exhaust plume are difficult to access during engine operation, particularly during flight; the possibility that monitoring equipment could interfere with the normal performance of the engine; the complexity of the physical and chemical processes occurring in the engine and exhaust plume; the very short lifetimes of some of the intermediate species produced; and the variation of the relevant physical and chemical processes with a range of ambient conditions, engine designs and operational factors. Despite these difficulties, some direct monitoring of aviation emissions has been undertaken and the plumes and wakes of aircraft have been sampled to determine the species they contain. For instance, recent research has focused on investigating the gaseous and particle emissions of aircraft using the ALFA aircraft plume analysis facility. In particular, high resolution mass spectrometry may be used to measure aircraft particle emissions, of which scientific knowledge is currently relatively limited (Lee 2009).

Indirect forms of monitoring may also be used to estimate aviation emissions. Those methods involve measuring ambient concentrations of key pollutants (such as NO_2 or SO_2), including the emissions due both to aviation and to non-aviation sources. A standardised methodology may then be used to estimate the proportion of those pollutants that may be attributed to aviation sources, an approach known as source apportionment. This method has the advantage that some key pollutants, including NO_2 and SO_2, are regulated by national legislation in many countries; hence standardised, regular monitoring of those species already occurs at many locations (see Chapter 4). However, this approach also has several disadvantages: not all aviation emissions are defined as key pollutants at a national level and, if they are, the key pollutants that are monitored (such as NO_2) may not correspond exactly to the substances emitted by aircraft (such as NO_x). Also, this approach requires that all the significant sources of pollution in a given area are known and that the transport processes responsible for dispersing substances in the atmosphere are well understood – which is often not the case. Furthermore, indirect monitoring and source apportionment may only be used at the local scale (in places where adequate air quality monitoring occurs); those techniques cannot provide much understanding of aviation emissions at the national, international or global scale. Nevertheless, many airports now undertake some form of air quality monitoring and detailed studies have also been undertaken to estimate airport-related emissions of key pollutants based on scientific methods (DfT 2006; Peace et al. 2006; Schürmann et al. 2007; Unique 2004).

Given the difficulties involved in monitoring aviation emissions directly, as well as the uncertainties and limitations associated with indirect monitoring and source apportionment methods, other approaches to characterising aviation emissions have focused on the use of modelling techniques. Emissions modelling can be used to characterise aircraft emissions at the global scale and several such estimates have been produced (Eyers et al. 2004; IPCC 1999; Kim et al. 2007; Lee 2004; Lee and Raper 2003; Lee et al. 2007; Rogers et al. 2002). In this context, modelling techniques are used to produce emission inventories of specified substances that are generated by the global aircraft fleet. Global emission inventories have been produced for current (or for early 1990s) aircraft emissions; scenarios have also been developed for the years 2015 and 2050 based on assumptions about the relationship between revenue passenger kilometres and gross domestic product (GDP), and about the pace of development of technology to improve fuel consumption and to reduce NO_x emissions (Lee 2004; Lee and Raper 2003; Peeters and Williams 2009). Global aircraft emission inventories generally produce three-dimensional (3D) gridded data and they necessarily use various simplifying assumptions. They contain several essential elements: (a) a database of aircraft movements; (b) a representation of the aircraft and engine types that constitute the global aircraft fleet; (c) a fuel-flow model; (d) a method for calculating emissions at typical cruising levels from fuel flow values; and (e) emission data for the aircraft landing and take-off (LTO) cycle (Lee and Raper 2003). By using 3D inventories of aircraft emissions together with a 3D chemical transport model of the global atmosphere (which simulates the way in which substances are transported by atmospheric processes), it is possible to determine the effect of aircraft emissions on the composition of the atmosphere (Lee and Raper 2003). Modelling techniques of this type have been used to demonstrate that global aircraft emissions are concentrated in the Northern Hemisphere at altitudes of between eight and 12 kilometres, where most civil flights occur (Lee 2004).

One important application of emissions models is to study the ways in which aircraft emissions may change in the future. Again, various studies have investigated this topic with a particular focus on two key emissions, CO_2 and NO_x (changes in aircraft fuel consumption are also necessarily modelled in such studies). The results indicate that aviation CO_2 emissions are increasing rapidly and could more than double by 2025 (Eyers et al. 2004; Lee 2004). By 2050, under some scenarios, aviation CO_2 emissions are projected to treble over 1992 levels, even when optimistic assumptions about technological improvements are made (Horton 2006; IPCC 2007;

Lee 2004; 2009; Owen and Lee 2006; Peeters and Williams 2009; Stern 2007; WRI 2005; Yamin and Depledge 2004; Figure 2.4). In the case of NO_x emissions, by 2015, emissions from civil aviation emissions (in the form of NO_2) and aircraft fuel consumption are forecast to almost double, reaching 3.53Tg N yr^{-1}; and, by 2050, NO_x emissions and fuel consumption may have increased by factors of approximately 2 to 6 over early 1990s levels (Lee 2004; Lee and Raper 2003). Given that these forecasts and projections cover relatively long time periods, the assumptions made in the models can exert a strong influence on the results, meaning that careful scrutiny of the validity of the simplifying assumptions is required. It is clear that assumptions about GDP growth are critical to the overall emissions levels projected for 2050, whilst other assumptions about the rate of technological progress have a second-order effect (Lee 2004; Lee and Raper 2003). Furthermore, global aviation emission inventories must be updated regularly as improved data and better understanding of atmospheric processes become available, although that task represents a major undertaking (Lee 2004; Peeters and Williams 2009).

As previously mentioned, aircraft emissions are significant because they cause various environmental impacts at the global and local scales; those impacts, and a range of ways in which they may be reduced, are discussed in more detail in the next two chapters. However, some general comments about the regulation of aviation emissions are included here. Aviation emissions are currently regulated through the ICAO engine certification process, which requires that engines meet relatively stringent standards for emissions of four pollutants (NO_x, HCs, CO and smoke) over the aircraft LTO cycle (ICAO 1981, 1995). In this context, 'smoke' is a crude measure of particle emissions, particularly of soot. Measurements of emissions from a limited number of new (or recently manufactured) engines are made using standardised methods, and the results are published in the ICAO Engine Emissions Databank. However, Lee (2004, 4) has acknowledged that the ICAO certification standards 'are manufacturing standards, not an in-service compliance regime'; hence those standards do not take account of the marked decline in engine performance that occurs over the course of an engine's service life due to the deterioration of its component parts and the accumulation of deposits on blade surfaces. Neither do they reflect the increased emissions generated over the service life of an aircraft due to the deterioration of its airframe, which leads to increased aerodynamic drag and higher rates of fuel consumption (Curran 2006; Eyers et al. 2004). Notably, the ICAO certification standards apply only to emissions over the LTO cycle and not at cruise altitudes (although technological improvements that limit emissions over the LTO cycle generally also reduce emissions during cruise; Lee 2004). More significant, however, is the fact that CO_2 emissions are not regulated by the ICAO certification process, since those certification standards were originally developed in response to concerns about local air pollution – and CO_2 has no effect on local air quality. Recently, in Kyoto Protocol signatory states, CO_2 emissions from aircraft on domestic flights within those countries have been accounted for, and reported, in the context of national carbon allowances. Similarly, CO_2 emissions from aircraft engaged in international flights in European airspace have been progressively brought within the scope of the EU ETS, thereby making polluters more accountable for their emissions. These initiatives are in their early stages, however, and they do not yet represent the direct regulation of CO_2 emissions from aircraft engines. Besides the ICAO certification standards, no other form of direct regulation of aircraft emissions exists.

Nevertheless, over several decades, significant improvements in aircraft fuel efficiency have been achieved in order to reduce operating costs (Figure 2.5). Those improvements have had the side-effect of reducing most specific aviation emissions; they have been so impressive that, of the pollutants regulated by ICAO, only NO_x remains a significant environmental and technological challenge (Lee 2004). Efforts are being made to achieve further improvements in the environmental performance of aircraft and, in particular, to reduce NO_x emissions, although that task involves a complex interaction with simultaneous efforts to improve fuel efficiency and to reduce CO_2 emissions (discussed previously). Nevertheless, the level of stringency of the ICAO certification standards is periodically reviewed and is occasionally increased (for new engines introduced to the global fleet) following negotiations and agreements within the ICAO Committee on Aviation Environmental Protection (CAEP). In February 2004, CAEP held its sixth meeting (CAEP 6) in Montreal and a revised standard for aircraft NO_x emissions (a reduction of 12 per cent compared to the previous standard) was agreed for new aircraft entering service from 2008. Again, it is worth emphasising that CO_2 emissions remain unregulated by ICAO.

The phrase 'reducing emissions' requires careful interpretation. Claims that emissions have been reduced often refer to specific emissions: the quantity of a substance emitted per unit of traffic carried (which may be expressed in terms of revenue passenger-kilometres or freight tonne-kilometres). Therefore, specific emissions reductions are relative to the total number of passengers or the total mass of freight transported, and also to the total distance flown, so they are measures of the efficiency with which air transport operates. However, as noted previously, and as will also be explored in the next section, the growth of air transport has been rapid and sustained and is projected to continue on a decadal timescale. Hence the total number of revenue passenger-kilometres (and freight tonne-kilometres) has increased dramatically – more rapidly than the rate at which emissions have decreased due to improvements in aviation technology and in operational procedures. Consequently, specific emissions have generally declined. On the other hand, absolute emissions have increased due to the vastly increased demand for air transport (Royal Academy of Engineering 2003). Therefore, although the air transport industry now operates with unprecedented efficiency, it also generates more emissions than at any time in its history. In assessing the overall environmental impacts of aviation, a key point is that the growth of air transport has outpaced any reductions in specific emissions that have been achieved due to improved technologies and operational procedures (Bows et al. 2009; IPCC 1999). Understanding the characteristics of aviation growth, therefore, is critical to understanding the overall environmental impacts of air transport. Aviation growth is explored in more detail in the next section.

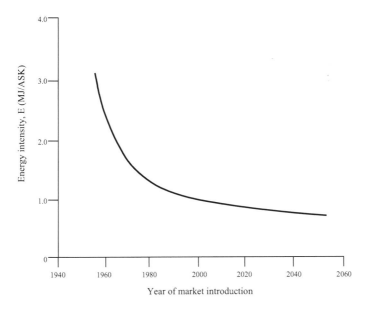

Figure 2.5   Improvements in aircraft fuel efficiency, c. 1960–2050

Source: Adapted from Peeters et al. (2009, 294)

Understanding Aviation Growth

Many commentators have remarked that aviation is growing rapidly; indeed, this is often one of the first observations to be made in contemporary analyses of air transport (Bishop and Grayling 2003; Bows et al. 2009; Button 2004; DfT 2003c; EUROCONTROL 2004; 2008; Gössling and Upham 2009; ICAO 2001; IPCC 1999; Lee et al. 2009; Upham 2003; Wright 1991). What does 'aviation growth' mean? The term may be used in various ways. Aviation growth can refer to the expansion of the industry itself, as measured in some standardised manner: by the number of flights or aircraft movements; by the number of units of traffic carried (revenue passenger-kilometres or freight tonne-kilometres); by the number of passengers passing through airports; by the number of aircraft or engines manufactured or sold; by the number and size of airports; by the number of employees working in the aviation industry; or by air transport revenue. Air passenger traffic is often expressed in terms of revenue passenger-kilometres (RPKs), defined as the number of fare-paying passengers multiplied by the distance that those passengers fly. Air freight traffic is typically measured in terms of revenue tonne-kilometres (RTKs) or freight tonne-kilometres (FTKs), defined as the mass of air cargo multiplied by the distance that the cargo is carried. Various other measures of the size of the aviation industry exist, including measures of the size and composition of the global fleet (including the total number of aircraft, the number of aircraft in service, the number of aircraft parked or retired, the number of new aircraft deliveries and the numbers of aircraft of specific sizes or types). In addition to measures focusing on aircraft numbers, growth can also be measured in terms of frequencies (the number of flights undertaken by airlines), the number of routes flown or the number of new routes opened. Other measures of aviation growth relate to load factors, meaning the number of fare-paying passengers as a proportion of the total number of available seats on aircraft. Load factors indicate how efficiently the available seats are used; in 2005, global airline load factors were reported to have reached record levels exceeding 76 per cent (Airbus 2007; Boeing 2006; King 2007). Finally, growth can be measured in terms of market values, which are usually adjusted to an index year and are expressed in terms of trillions of US dollars.

All of the measures listed above indicate changes in the magnitude of the output of the industry; they do not reveal the quality of that expansion, such as changes in the quality of the service that passengers receive, or in the speed and reliability of air transport services. Hence aviation growth may also refer to qualitative improvements in the air transport industry, involving fundamental changes in the structure and organisation of the industry and in the ways in which services are delivered (although such qualitative improvements may be better characterised as the 'development' or 'evolution' rather than the 'growth' of the industry). Aviation growth does not always refer to the scale of the operation of the air transport industry *per se*, however; it may also refer to increasing demand for air transport. Many observations about the growth of air transport actually focus on the growth of demand for air transport services (Airbus 2007; Boeing 2008; Bows et al. 2009; Button 2004; Cairns and Newson 2006; DfT 2003c; EUROCONTROL 2004; 2008; Graham 2001). A further distinction that may be drawn is that the growth of demand for air transport may be unconstrained (meaning the demand that would occur if the capacity of the air transport system were infinite) or constrained (by existing or expected capacity limits). Different conclusions about the nature and scale of aviation growth may be reached depending upon which of these various interpretations is used.

The demand for air transport and the actual scale of air transport operations are closely related – and both are increasing dramatically. Nevertheless, that relationship is not straightforward. Demand generally far exceeds the capacity of the air transport system, meaning that significant demand is not realised due to capacity constraints (see Auerbach and Koch 2007; DfT 2003c; Forsyth 2007; Givoni and Rietveld 2009; Graham and Guyer 1999; Hanks 2006; Madas and Zografos 2008). Capacity constraints may be complex and can occur throughout the air transport system; the provision of air transport services may be limited by insufficient capacity of the aircraft, airspace, air traffic management (ATM) system, runways, taxiways, aprons, stands, terminals, ground handling services or the surface transport routes to and from airports. Therefore, a tension exists between demand and capacity, and policy debates have focused on the extent to which aviation capacity should be increased to meet the expected demand – the so-called 'predict and provide' approach to aviation infrastructure (Cairns and Newson 2006; RCEP 2002). Accurately forecasting demand for air transport is difficult, however, since demand is affected by price signals that in turn respond to the supply of air transport services (and hence to capacity). Furthermore, Button (2004) has acknowledged that demand for air transport is a derived demand: most people use air transport to meet some other need, rather than as an end in itself, so that demand for air transport also varies with external shocks (Airbus 2007; Bows et al. 2009; Njegovan 2006). Nevertheless, some authors have drawn attention to the apparently irreconcilable tensions resulting from simultaneous attempts to service the growing demand for air transport and to promote environmental sustainability (Bows et al. 2006, 2009; Cairns and Newson 2006; Graham and Guyer 1999). Therefore, concerns about the environmental impacts of air transport now represent an additional constraint of the growth of the industry – a concept that is referred to as limited environmental capacity (Graham 2001; Upham 2001; Upham et al. 2003; 2004).

Reconstructions of the historical growth of air transport, based on air passenger and freight traffic data, indicate that aviation has expanded rapidly since 1960 as the global economy has grown (Graham 1995; Figure 2.6). Between 1960 and 1999, air passenger traffic (expressed in terms of RPKs) increased dramatically, at almost nine per cent per year, a factor of 2.4 times the rate of increase of global average GDP. Air freight – the majority of which has conventionally been carried on passenger aircraft (although that trend is changing) – also increased over the same time period at an average rate of around six per cent per year. Those very high growth rates equate to a 23-fold increase in the output of the air transport industry (measured in terms of tonne-kilometres performed) since 1960 (Bailey 2007; IPCC 1999). By 1997, the rate of growth of air passenger traffic had slowed to around five per cent as the industry matured, and periodic downturns have occurred over relatively short timescales in response to particular events, including the terrorist attacks in the USA in September 2001, the outbreak of severe acute respiratory syndrome (SARS) in China and parts of Southeast Asia in 2002–2003, and the global financial crisis of 2008–2010. Nevertheless, growth rates have tended to recover rapidly after such downturns (Lee et al. 2009). Airbus (2006) reported that, in 2004, air traffic grew at an average annual rate of fourteen per cent – greater than at any time in the preceding 25 years – and that the average annual growth rate remained high, at seven per cent, in 2005. Overall, long-term global aviation growth rates of around five per cent per year have been sustained – well in excess of the rate of increase of GDP (Lee 2004). For the two decades from 1985 to 2005, Boeing (2008) reported that the average annual passenger traffic growth rate was 4.8 per cent and the average annual air cargo growth rate was 6.3 per cent, both of which exceeded the global average annual economic growth rate of 2.9 per cent.

What has driven this sustained, rapid growth of air transport? Aviation growth has been reinforced by three important, interrelated trends: economic growth, globalisation and the growth of tourism (Debbage 1994; Forsyth 2008; Graham 2008; Janelle and Beuthe 1997; Mayor and Tol 2010; Pels 2008). Although the very rapid expansion of the air transport industry has greatly outpaced the growth of the world economy, both air transport growth and economic growth are closely related. Historically,

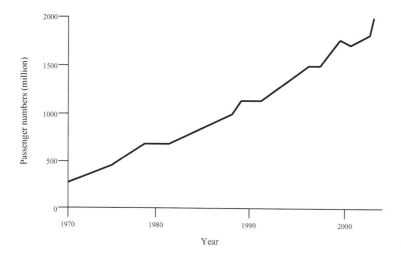

Figure 2.6   The growth of air transport expressed in terms of passenger numbers, 1970–2005

Source: Adapted from Bows et al. (2009, 11)

the relationship between GDP growth and transport growth has been strong; aviation has expanded rapidly with global economic growth, with increasing GDP being responsible for approximately two-thirds of the growth in air transport (IPCC 1999). In particular, the growth of air transport has been stimulated by the globalisation of manufacturing industry, commerce and finance, and air transport has become a significant economic generator and an integral part of business, commerce and trade (Pels 2008; Rogers et al. 2002). Globalisation is now acknowledged to be a complex process that is responsible for profound economic, environmental and social transformations worldwide. Communications between some places are now almost instantaneous; global transportation is commonplace; the world economy has become increasingly integrated; the influence of multinational organisations has expanded; and the autonomy of most nation-states has diminished (Hettne 2008). Air transport has been important in facilitating globalisation, and globalisation has in turn increased demand for air travel (Goetz and Graham 2004; Janelle and Beuthe 1997; Young 1997). The growth of demand for air transport has also been stimulated by the rising demand for tourism, an industry that is highly dependent upon the availability of rapid, long-distance air services (Abeyratne 1999; Becken 2007; Bieger and Wittmer 2006; Daley et al. 2008; Dubois and Ceron 2006; Gössling 2002; Gössling and Peeters

2007; Gössling et al. 2005; Graham 2008; Graham et al. 2008; May 2002; Mayor and Tol 2010; Patterson et al. 2006; Peeters et al. 2006; Peeters and Schouten 2006; Shaw and Thomas 2006). At the global scale, tourism, like aviation, is an important economic driver and is projected to expand rapidly: at an average rate of around four per cent per year until at least 2020. Projections by the United Nations World Tourism Organization (UNWTO) indicate that international tourist arrivals (ITAs) will double between 2005 and 2020 and are expected to reach 1.6 billion by the latter year – a trend that has important implications for the scale and distribution of aviation growth (UNWTO 2008).

Given the strong forces driving its growth, the air transport industry is expected to continue to expand in the future. Various predictions of the growth of the industry have been made for periods covering decadal timescales. Many studies indicate that sustained growth in demand for air transport of around five per cent per year is likely to continue until at least 2030 (Bailey 2007; Bieger et al. 2007; Bows et al. 2005; 2006; Button 2004; DfT 2003c; 2004; IPCC 1999). Indeed, some authors have argued that sustained growth in demand will occur over a much longer period, to 2100 (Mayor and Tol 2010). Total aviation fuel use (including fuel used by passenger, freight and military aircraft) has been projected to increase by the smaller figure of three per cent per year between 1990 and 2015, due to expected improvements in the fuel efficiency of aircraft (IPCC 1999). Of particular importance in understanding aviation growth are the influential market forecasts produced by aircraft and engine manufacturers, such as the alternative visions for air transport growth published by Airbus and Boeing (Airbus 2007; Boeing 2008). Commercial market forecasts are generally based on analyses of the domestic, regional and intercontinental passenger sub-markets, including the flows between and within regions, and also of directional air freight sub-markets; the results are published in the form of detailed passenger, freight and fleet forecasts. In producing those forecasts, traffic flows are analysed at a relatively detailed level and are aggregated into predictions for the main travel regions of the world. The analyses are undertaken by organisations such as the Global Insight Forecasting Group, using projected economic growth rates in combination with other indicators, such as oil prices. Air freight forecasts incorporate assessments of previous multidirectional import and export levels and are categorised by country and by item (Bows et al. 2009). The fact that these forecasts of the growth of air transport are based on assessments of economic growth emphasises the strong correlation between air transport growth and the strength of the world economy (Rogers et al. 2002).

Commercial market forecasts predict that air transport will grow at an average annual rate of around five per cent for the next twenty years. Airbus (2007) has stated that global passenger traffic is expected to increase by 4.9 per cent per year, and freight traffic by 5.8 per cent, over the period 2007–2026, and that air traffic will double over the fifteen-year period 2007–2021. Boeing (2008) has forecast the average annual growth rates to be 5.0 per cent for passenger travel and 5.8 per cent for freight. Globally, passenger traffic is forecast to increase from 4.5 trillion RPKs in 2007 to 12 trillion RPKs in 2027 (Boeing 2008). Comparable growth forecasts have been published by the engine manufacturer Rolls-Royce, which has stated that the global fleet of airliners is expected to double over the twenty year period 2007–2026, although the average rate of global traffic growth is predicted to be faster still (at 4.9 per cent per year) due to increases in frequencies, load factors and aircraft size (Rolls-Royce 2007; see also Bows et al. 2009). Additionally, Rolls-Royce (2007) has acknowledged that strong growth is expected to occur in another part of the aviation sector: the manufacture of business jets (see also Budd and Graham 2009). The commercial market forecasts predict that aviation growth will continue rising dramatically in Asia, especially in China and India which are on course to become the world's largest consumer markets by 2033 (Lei 2008; O'Connell 2008; see also Rengaraju and Arasan, 1992). Boeing (2008) has argued that the fastest growing economies will prompt the formation of a more geographically-balanced market. However, Boeing (2008) has also acknowledged that relatively slow growth in the largest, established markets may yet generate more traffic than rapid growth in many of the smaller markets.

The commercial market forecasts differ in their view of the distribution and characteristics of future aviation growth. In particular, Airbus and Boeing have adopted 'consolidation' and 'fragmentation' perspectives towards aviation growth, respectively. The consolidation model predicts that the growth of the industry will be focused on major hub airports served by increasing numbers of very large aircraft, such as the Airbus A380. The hub airports, in turn, are already capacity-constrained (or nearly so) and will require considerable infrastructure development to accommodate both increased passenger throughput and larger aircraft (Airbus 2007). Networks of spoke routes radiating from the hub airports would serve peripheral airports using smaller aircraft. One implication of the consolidation model is that aircraft environmental impacts could become increasingly concentrated at and around the hub airports (Lu and Morrell 2006); in some cases, emissions are expected to increase more rapidly than traffic due to the use of larger, heavier aircraft. Rogers et al. (2002) have argued that the increasing absolute environmental impacts associated with the growth of hub airports may be partly offset by technological improvements, although further research is required to fully understand those impacts. In contrast, the fragmentation model of the growth of air transport predicts that increasing numbers of passengers will fly between individual city-pairs without the need to travel via a third, hub airport (Boeing 2008). The fragmentation model implies a large increase in the number of aircraft required to service the many extra routes that would be flown; however, it also implies shorter average flight distances and thus, potentially, reduced emissions (Morrell and Lu 2006). This view has influenced the development of medium-sized aircraft (such as the Boeing 787) that are designed to connect a large number of city-pairs. The fragmentation model also implies that more airports will be required, in more cities, in order to maximise the choice of routes offered to passengers. Rogers et al. (2002) have acknowledged that the fragmentation model would mean that environmental impacts – especially noise and emissions – from individual aircraft would be smaller than those produced by the very large aircraft envisaged in the consolidation model, but that those smaller impacts would be offset by a considerable increase in the total number of aircraft operating. Furthermore, the development of more city-pair routes would spread aircraft environmental impacts to areas that had previously been relatively untouched by those impacts. The consolidation and fragmentation models represent alternative views of the growth of air transport at the global scale (King 2007; Mason 2007; Swan 2007). It remains unclear which of these divergent views will be the more dominant.

In addition to these forecasts, authoritative scenarios of possible global aviation growth have been published by the IPCC for the longer period 1990–2050; they were developed by the ICAO Forecasting and Economic Sub-Group (FESG) of CAEP. Those future global aircraft scenarios, known as the FESG scenarios, were informed by another range of scenarios, termed IS92a-f, which considered future greenhouse gas and aerosol precursor emissions for the period 1990–2100 (IPCC 1992). The IS92 scenarios were based on assumptions about population and economic growth, land use, technological changes, energy availability and fuel mix; in that range of scenarios, IS92a represented a 'midrange' growth scenario from which the Fa1 'reference scenario' of global aviation growth was developed (IPCC 1999). It is important to emphasise that the FESG scenarios adopted a broader approach than the commercial market forecasts discussed previously. The development of the FESG scenarios was informed by several models of air traffic demand and took account of the market maturity concept: the idea that the very high historical rates of air transport growth – greatly exceeding the rate of world economic growth – are unlikely to continue indefinitely. The market maturity concept suggests that the aviation growth rate is likely to approach the rate of GDP growth as air transport markets become mature. This approach is itself based on several assumptions: that the global air transport market can be treated as a single, maturing market comprising regional markets at varying stages of maturity; that historical trends in the growth of air transport and of the world economy provide a reasonable basis for estimating future growth rates; that the growth of business and personal air travel sectors can be considered together; that air transport growth is driven by GDP growth; that fuel prices and availability will continue to permit the expansion of the industry; that air transport growth will be unconstrained by regulations, technology or infrastructure; and that the industry will not be out-competed by other travel modes (such as high-speed rail) or technologies (such as telecommunications). Therefore, the FESG scenarios assume that the air transport industry will continue to operate along broadly similar lines to the present industry and that no radical social or technological changes will disrupt the current pattern of regional markets (IPCC 1999). One key observation to emerge from the work based on the FESG scenarios is that, by 2050, air passenger traffic is expected to have increased five-fold over 1995 levels (IPCC 1999).

In the discussion presented above, attention is drawn to the distinction between forecasts (shorter-term predictions of what is actually expected to happen) and scenarios (longer-term possibilities that may or may not eventuate, which are based on various assumptions about economic and population growth, technological development, government policies and operating practices). Forecasts of the growth of air transport up to 2025 can be made with a relatively high degree of confidence due to the long development times required to introduce new aircraft, airport infrastructure and aerospace technologies; the air transport industry of 2025 is likely to operate along broadly similar lines to that of the present day (for an alternative view, see Riddington 2006). However, predictions of air transport growth beyond 2025 are characterised by greater uncertainties; therefore, instead of using growth forecasts for the more distant future, an alternative approach – one based on a range of scenarios – is necessary. The FESG scenarios share several common assumptions: they all assume that technological improvements will occur, leading to reduced emissions per RPK, and that the optimal use of airspace – in other words, ideal ATM – will have become a reality by 2050 (IPCC 1999). Hence the FESG scenarios represent a range of possible trajectories of growth for aviation and they have also been used to inform global circulation models. Work continues in order to update and improve the scenarios used for projecting the long-term growth of air transport (see Bows et al. 2009).

Recently, improved scenarios have been used for the purpose of understanding aviation emissions. Since the assumptions underlying any range of scenarios are of critical importance for the validity of the output, those assumptions require periodic re-evaluation. The original IS92 scenarios were reviewed in 1995 in the light of improved scientific understanding of climate change and a new group of emissions scenarios was developed (IPCC 2000). Those updated scenarios, known as the IPCC Special Report Emissions Scenarios (SRES), were based on more accurate emission baselines and improved information about economic restructuring, worldwide. For instance, by that time, new insights into the carbon intensity of energy supply, the income gap between developed and developing countries, and sulphur emissions had become available (IPCC 2000). The SRES scenarios adopted a new approach, taking into account different rates and characteristics of technological change and considering a wider range of economic development pathways. Four narrative storylines were constructed to represent various possible demographic, social, economic, technological and environmental developments. Each SRES scenario formed a specific, quantitative interpretation of one of these storylines, and forty SRES scenarios were developed, which were categorised in six 'scenario groups' (IPCC 2000). Those scenarios demonstrate that different socio-economic assumptions – about demographic, social, economic and technological changes – result in different predictions of future levels of greenhouse gases and aerosols. Thus the SRES scenarios emphasise the strength of the link between alternative development paths, including adaptation and mitigation responses, and their associated levels of atmospheric emissions (IPCC 2001). Another important difference between the SRES scenarios and the IS92 scenarios is that the former contain lower projected SO_2 emissions, with the result that the temperature rises associated with the various SRES scenarios – and the range of those temperature rises – are greater than in the IS92 scenarios (IPCC 2001). Such issues illustrate the importance of using up-to-date scenarios, based on scientifically-defensible assumptions, in modelling the growth of aviation and its associated environmental impacts (see Mayor and Tol 2010).

Several other indications of the growth of air transport at a global scale have been produced. These include shorter-term forecasts, generally covering the period to 2015, issued by the UK Department of Trade and Industry (DTI), the European Civil Aviation Conference (ECAC)

Abatement of Nuisances Caused by Air Transport (ANCAT) and EC Emissions Inventory Database Group (EIDG), the Deutsches Zentrum für Luft- und Raumfahrt (DLR) and the Dutch Aviation Emissions and Evaluation of Reduction Options (AERO) project. All of those forecasts were based on broadly similar approaches (IPCC 1999). More recently, a growth forecast has been made for the year 2025 as part of the AERO2k global aviation emissions inventories study, although that forecast was based on air traffic growth data provided by Airbus (and other forecasts) and it reflects the commercial market forecast values (Eyers et al. 2004). Another forecast of aviation growth, for the year 2030, was produced for the DTI to facilitate assessments of the projected increase in aviation CO_2 emissions; that forecast was based on the FESG scenarios discussed previously and on the AERO2k database (Horton 2006). The UK Department for Transport (DfT) has produced national-level forecasts of the growth of air transport to 2030 (DfT 2003c; Riddington 2006). In addition to the FESG scenarios, various other long-term scenarios have been developed. One such projection has been made by the DTI for the period to 2050, which produced an overall estimate of over 18×10^9 RPKs by 2050: a value that occupies an intermediate position compared with the range of the FESG scenarios. Longer-term projections, to the year 2100, have been produced by the Environmental Defense Fund (EDF); those projections are based on the IS92 scenarios discussed above, although they also incorporate assumptions about the use of very low NO_x technology. Since 2001, the FAA has developed the System for assessing Aviation's Global Emissions (SAGE), based on annual global inventories of commercial aircraft fuel burn and emissions, which forms the basis for scenario modelling (Kim et al. 2007; Lee et al. 2007).

Other projections have been developed by the Massachusetts Institute of Technology (MIT) and by the World Wide Fund for Nature (WWF), with the latter covering the period to 2041 (IPCC 1999). The MIT and WWF scenarios included assumptions relating to a wide range of policy and operational choices, such as the achievement of 100 per cent load factors and the introduction of 100 per cent fuel taxation; nevertheless, the WWF scenarios produced an average annual growth rate (5.2 per cent) similar to that predicted by the commercial market forecasts, and they also projected potential increases in demand for air transport ranging from a factor of three under a strict demand management scenario to a factor of more than twelve under a 'business-as-usual' scenario (IPCC 1999). The Global Scenario Group (GSG) of the Stockholm Environment Institute has produced scenarios of global changes for the years 1995, 2025 and 2050 incorporating the main modes of transport and highlighting the implications for mobility (Timms et al. 2005). A transport model has also been produced by the Tyndall Centre as part of its World Transport Scenarios Project; that model is based on the IPCC SRES scenarios and includes the main transport modes for the years 1995, 2025, 2050 and 2100 (Timms et al. 2005). For the EU, the Tyndall Centre has produced four scenarios covering the period 2012–2051, three of which are constrained by an EU commitment to maintain CO_2 concentrations within 450 ppmv, and all of which contain assumptions about aviation emission growth rates (Anderson et al. 2005; Bows et al. 2009). At the national level, the Tyndall Centre has developed various sets of scenarios for the UK, which are constrained by the need to meet the UK's 60 per cent CO_2 reduction target and which also include a range of assumptions about the growth of aviation emissions (Anderson et al. 2005; 2008; Bows et al., 2005; 2009; Mander et al. 2008).

The longer-term projections of air transport growth discussed above contain many uncertainties. There are many reasons why this is inevitably the case: appropriate technological developments are difficult to forecast; patterns of demand may vary; and economic and social development trends are extremely uncertain. The air transport industry is also likely to face significant obstacles to growth as a result of constraints due to limited operational and environmental capacity of airports, congested airspace, ATM inefficiencies, environmental regulations and even shortages of kerosene (Allen 1999; Andrieu 1993). All of these potential constraints could substantially curtail the growth of the industry. The availability of kerosene, in particular, could represent a limiting factor for aviation growth since, at the global scale, conventional oil resources could be significantly depleted by 2050 (IPCC 1999). As a result of these uncertainties, forecasts of air transport growth beyond 2025 may be highly inaccurate, meaning that a scenario-based approach – in which the various underlying assumptions can be evaluated – is more appropriate for the development of projections to 2050 (IPCC 1999). Whilst the air transport market may be reaching maturity in some regions of the world, and thus may operate along more predictable lines in those regions, this is not necessarily the case worldwide. Furthermore, some regional air transport markets will probably develop and mature much more rapidly than the US domestic air transport market, which reached maturity over a period of approximately seventy years (Airbus 2007; Boeing 2008; IPCC 1999). Whilst none of the long-term scenarios for the growth of air transport mentioned above is impossible, the more extreme high- and low-growth scenarios are improbable due to infrastructure constraints and insufficiently radical technological developments, respectively. Hence the growth of air transport at the global scale will probably resemble one of the intermediate scenarios. Crucially, whichever scenario is considered, the projected growth of air transport will have significant environmental implications, as subsequent chapters will show.

Summary

This chapter has explored two key themes – aviation emissions and aviation growth – and in so doing has provided a context for the discussion of environmental impacts that follows. Aircraft emit gases and particles directly into the atmosphere. The major emissions from aircraft are the greenhouse gases carbon dioxide (CO_2) and water vapour (H_2O); other emissions include nitrogen oxides (NO_x), sulphur oxides (SO_x), hydrocarbons (HCs), carbon monoxide (CO) and particles (including soot and sulphate particles), which have a range of environmental effects. Airports, as well as aircraft, produce emissions due to the combustion of fossil fuels in a variety of airport-related processes. Although technological and operational progress has been made in reducing aviation emissions, and the environmental performance of aircraft and engines has improved dramatically over many decades, the absolute emissions of air transport are nonetheless increasing because the growth of the air transport industry has outpaced those improvements. Whilst the emissions from aircraft and airports – especially NO_x and particle emissions – are responsible for localised air pollution in the vicinity of airports, the most significant concerns about aviation emissions relate to their climate impacts, including the unique effects of aircraft due to their operation at cruising levels. During cruise, aircraft emissions are deposited directly into a climatically-sensitive region in the upper troposphere and lower stratosphere where they form the predominant anthropogenic emissions and make a significant – and increasing – contribution to climate change.

Aviation growth has been largely driven by economic growth, globalisation and the growth of tourism. Consequently, since 1960, the aviation industry has experienced consistently high growth rates – of around 5 per cent per year – which have far exceeded the rate of global economic growth. This chapter has considered various aspects of aviation growth, including the commercial market forecasts and the authoritative scenarios that provide insights into the ways in which aviation and emissions are likely to continue to grow strongly over decadal timeframes. The growth of air transport is rapidly transforming aviation from being a relatively minor polluter into a highly significant source of radiative forcing of climate (in addition to its other environmental impacts). Even if a conservative estimate of the growth of aviation emissions is used, CO_2 emissions from air transport are expected to quadruple over the period 1990–2050, and some forecasts suggest that CO_2 emissions from air transport could increase ten-fold over that period (Cairns and Newson 2006). Depending on the scenario used, the air transport industry could consume most – or even all – of national carbon budget allowances under the Kyoto Protocol or its successor agreement (Bows et al. 2009). The environmental impacts of air transport – discussed in the next three chapters – should therefore be interpreted in the context of the very rapid growth of aviation and of aircraft emissions. Given the long timescales required to develop and introduce new aviation technologies, and the long in-service lifetimes of aircraft, the task of reducing the projected environmental impacts of air transport has already become urgent.

3 Air Transport and Climate Change

Introduction

The issue of climate change has become prominent in recent scientific and policy debates and now represents one of the dominant global environmental issues (Adams 2009; Houghton 2009; IPCC 2007; Stern 2007). Popular and scientific concerns have been expressed about the effects and impacts of climate change, which are projected to become increasingly profound and widespread. As a result, concerted efforts at all levels are required to adapt to, and to mitigate, those impacts (IPCC 2007; Stern 2007). Air transport is currently a small but nonetheless significant contributor to climate change, mainly as a result of the various, interrelated effects of aircraft emissions (for an overview, see Chapter 2). However, given the projected sustained rapid growth of the air transport industry of around five per cent per year, aircraft emissions are expected to constitute a substantial proportion of global greenhouse gas emissions by 2050 (Bows et al. 2009). The task of reducing the impact of air transport on climate represents a major challenge. The fact that the industry is international in its scope, is technologically mature and is characterised by long lead-in times and long in-service times means that achieving emissions reductions will be extremely difficult (Gössling and Upham 2009; Lee 2004: Peeters et al. 2009). Besides the effects of aircraft emissions, other aviation-related impacts on climate occur due to the operation of airports and their ancillary services, and the prospect of reducing the emissions generated by those activities may be greater. Despite the challenges involved, there exists a growing determination to reduce the overall impact of aviation on climate, as the G8 Gleneagles climate statement, signed on 8 July 2005, attests (Forster et al. 2006; Upham and Gössling 2009).

This chapter begins with an overview of climate change in general, providing a brief explanation of the effects and impacts of climate change and the necessary responses (adaptation and mitigation). Some key concepts – such as that of radiative forcing – are explained. Thereafter, the chapter turns to the contribution of air transport to climate change. The various direct and indirect effects of aircraft emissions are considered; those effects include the emission of greenhouse gases such as carbon dioxide (CO_2) as well as the influence of other substances. Whilst an overview of aviation emissions has already been provided in Chapter 2, the account presented below focuses on the specific effects of those emissions on climate; therefore, the material contained in this chapter expands on the previous account. A critical issue to emerge in recent debates is that the distinction between the influence of CO_2 and the other, 'non-CO_2' effects of aviation is important in evaluating the overall impact of air transport on climate; that issue is highlighted in this chapter. Whilst this chapter focuses primarily on emissions from aircraft (which represent the dominant impact of aviation on climate), airports and their ancillary services also have some important effects on climate, and those effects are also considered. Next, in this chapter, various responses to the challenge of reducing the impacts of air transport on climate are considered. Those responses include a range of technological and operational approaches: the former seek to achieve improvements in airframes, engines and fuel, whilst the latter involve a wide range of adjustments to the ways in which aircraft are flown, such as revised cruising levels. This chapter also considers a range of policy options for reducing the impacts of air transport on climate, including regulatory, market-based and voluntary approaches. Several important – yet controversial – policy options, such as the use of emissions trading schemes and carbon offsetting schemes, are considered in this chapter.

Climate Change

Climate change is not a new environmental issue; in 1896, Svante Arrhenius suggested that the CO_2 released by human activities could increase global temperature (Arrhenius 1896; Houghton 2009; Seinfeld and Pandis 2006; Stern 2007). However, climate change – a term that encompasses increasing global temperature and its associated impacts – has gained much greater prominence in recent scientific and policy debates (Adams 2009; Houghton 2009; IPCC 2007; Stern 2007). Climate change requires careful definition, since climate varies naturally over all temporal and spatial scales (Seinfeld and Pandis 2006). 'Climate' may be defined as the condition of the atmosphere over many years: 'the mean behaviour of the weather over some appropriate averaging time', which is conventionally taken to be 30 years (Seinfeld and Pandis 2006, 4, 1026). 'Climate change' may be defined in various ways. The Intergovernmental Panel on Climate Change (IPCC) has defined climate change in the following terms:

> Climate change [...] refers to a change in the state of the climate that can be identified (e.g. using statistical tests) by changes in the mean and/or the variability of its properties, and that persists for an extended period, typically decades or longer. It refers to any change in climate over time, whether due to natural variability or as a result of human activity. (IPCC 2007, 30)

In this definition, the IPCC acknowledged that climate change has both natural and human causes; climate change may result from natural processes, such as variations in the Earth's orbit, as well as from anthropogenic changes, such as alterations of the composition of the atmosphere or of land use patterns (IPCC 2007; Seinfeld and Pandis 2006).

In contrast, the United Nations Framework Convention on Climate Change (UNFCCC), in its Article 1, adopted a narrower view, defining climate change as: 'a change of climate which is attributed directly or indirectly to human activity that alters the composition of the global atmosphere and which is in addition to natural climate variability observed over comparable time periods.' (UN 1992, 3)

Therefore, the UNFCCC makes a distinction between 'climate change', which is regarded as a purely anthropogenic phenomenon caused by human activities altering the composition of the atmosphere, and 'climate variability', which is attributed to natural causes (IPCC 2007; UN 1992). Given its focus on curbing greenhouse gas emissions from human activities, it is unsurprising that the UNFCCC is concerned only with anthropogenic effects on climate. In IPCC usage, however, climate variability refers to variations in the means, standard deviations and ranges of climate parameters on all spatial and temporal scales beyond those of individual weather events, although those variations do not necessarily amount to a trend over an extended period. Climate variability – like climate change – may have both natural and anthropogenic causes (IPCC 2007). Overall, the IPCC's (2007) definition of climate change implies the existence of more consistent trends in climate variables beyond the periodic fluctuations that constitute climate variability. Such distinctions suggest that considerable care is required in defining and debating climate change.

Anthropogenic climate change occurs as a result of emissions of greenhouse gases, emissions of aerosols and changes in land cover.

Greenhouse gases released by various human activities – especially by the combustion of fossil fuels – absorb outgoing infrared radiation from the Earth's surface and help to retain heat in the atmosphere. The greenhouse gases covered by the Kyoto Protocol are carbon dioxide (CO_2), methane (CH_4), nitrous oxide (N_2O), hydrofluorocarbons (HFCs), perfluorocarbons (PFCs) and sulphur hexafluoride (SF_6) (UN 1998). The most significant anthropogenic greenhouse gas is CO_2 due to both its relative abundance and its long atmospheric lifetime; the release of anthropogenic CO_2 is projected to result in climate change on the millennial timescale and may affect the climate system for hundreds of thousands of years (Archer 2005; Houghton 2009; IPCC 2007; Lenton et al. 2006). Greenhouse gases vary in their effectiveness to retain heat in the atmosphere; consequently, they may be characterised in terms of their global warming potential (GWP), a measure of the total amount of CO_2 that would produce an equivalent warming over a given period as the greenhouse gas in question (Houghton 2009; Stern 2007). The total effect of greenhouse gases is partially concealed by the effect of aerosols (such as soot and sulphate particles; see Chapter 2), which scatter and absorb incoming solar radiation and which enhance cloud formation, resulting in a cooling effect on the atmosphere. Changes in land cover are also significant for several reasons: the reflectivity ('albedo') of the Earth's surface is modified; the capacity of vegetation to absorb CO_2 from the atmosphere is reduced by deforestation; and land is converted to uses that release greenhouse gases. Overall, the anthropogenic forcing of climate is dominated by the increase in greenhouse gas emissions, resulting in a net warming of the atmosphere.

Human and natural factors therefore influence climate by altering the radiative properties of the atmosphere by which energy is scattered, absorbed and re-emitted. Changes in those factors are known as radiative forcing, which is a measure of the importance of a particular climate change mechanism. Radiative forcing expresses a perturbation of the energy balance of the Earth-atmosphere system in watts per square metre (W m^{-2}); positive values indicate warming, while negative values imply cooling (Houghton 2009; IPCC 1999). The radiative forcing due to various natural events and human activities may be quantified, and the global mean radiative forcing exhibits an approximately linear relationship with the global mean surface temperature change (Lee 2004; Lee and Raper 2003; Seinfeld and Pandis 2006). The response of the climate system to sustained radiative forcing is known as the equilibrium climate sensitivity, which is defined as the equilibrium global average surface warming following a doubling of CO_2 concentration (IPCC 2007). The precise climate response to any given radiative forcing is complicated by the existence of feedbacks in the climate system which may amplify or dampen that response. For instance, rising global average temperature results in an increase in water vapour in the troposphere; since water vapour is a greenhouse gas, its increasing concentration will tend to exacerbate the temperature rise. Another feedback associated with rising temperature is the reduced capacity of land and ocean sinks to absorb CO_2, with the result that anthropogenic CO_2 tends to accumulate in the atmosphere more rapidly (IPCC 2007).

Climate change has many dimensions: variations in solar radiation, atmospheric circulation, sea level, ocean currents, clouds, rainfall, snow and ice, which are interconnected in complex ways. Such changes constitute the overall climate response to various sources of radiative forcing. Significantly, climate change involves changes not only in the mean values of those variables but also in their variances (Seinfeld and Pandis 2006). Climate change is characterised primarily by an increase in global annual mean surface temperature. The warming of the climate system is now acknowledged to be unequivocal: global average air and ocean temperature rises have been observed; widespread melting of snow and ice has occurred; and global average sea level rises have been recorded (IPCC 2007; Seinfeld and Pandis 2006). A variety of related effects has been observed, including the thermal expansion of the oceans and reduced snow and ice cover, which have in turn contributed to the rise in sea level. Rising temperature has also resulted in changes in the frequency and amount of precipitation, with some areas receiving significantly more rainfall and others experiencing declining rainfall and more frequent and severe drought. The occurrence and severity of extreme weather events have also changed; cold events have become rarer over most land areas whilst heat waves have become more prevalent. The occurrence of heavy precipitation events has generally increased, as has the incidence of extreme high sea level events. Intense tropical cyclone activity has also increased – an effect that is particularly apparent in the North Atlantic area (IPCC 2007; Stern 2007). The global temperature rise and its associated changes in sea level, rainfall, snow, ice and extreme weather events have already had a wide range of impacts on natural and human systems (IPCC 2007; Seinfeld and Pandis 2006).

In addition to the effects of climate change already observed, profound consequences are expected to occur in the future. A range of possible future conditions has been investigated by means of emissions scenarios which incorporate various projected changes in population, economic growth and technological development (IPCC 2000; 2007; see Chapter 2). The use of a range of emissions scenarios indicates that significant anthropogenic warming and sea level will continue for at least a century, even if emissions of greenhouse gases are severely curtailed (Houghton 2009; IPCC 2007). The long timescales involved in the climate response are due to a variety of factors, including the long residence time of CO_2 in the atmosphere and the slow rates of some atmospheric and oceanic transport processes. The impacts of climate change are expected to affect many systems, sectors and regions. The resilience of many ecosystems is likely to be exceeded, with many plant and animal species facing extinction. Profound changes in ecosystems and in the geographical ranges of species are expected to occur, with implications for biodiversity and for food supply. Coasts will be exposed to greater risks, including coastal erosion and saltwater inundation, and many coastal areas and their populations will be vulnerable to flooding. Many people will experience deteriorating health or mortality due to malnutrition, extreme weather events and changing patterns of disease. Climate change is expected to exacerbate stresses on water resources, with many semi-arid areas becoming drier, drought-affected areas spreading, demand for irrigation water increasing, freshwater species and ecosystems being subjected to greater stresses, and water quality declining (Houghton 2009; IPCC 2007; Stern 2007).

Climate change therefore presents an immense challenge to human activities since it threatens to disrupt economics, societies and environments in complex and interconnected ways world-wide. How should societies respond to this challenge? The human response to climate change takes two main forms: adaptation and mitigation. Adaptation involves societies modifying their activities in order to cope with those impacts of climate change that are already inevitable due to the long timescales over which the climate system adjusts to perturbations. However, in the long term, adaptation is not a sufficient response to the impacts of climate change, many of which are projected to increase with time; hence mitigation strategies are also required. Mitigation refers to the specific actions that societies can take to limit the magnitude and rate of climate change, primarily by reducing emissions of greenhouse gases. Adaptation and mitigation can reduce vulnerability to climate change impacts, both in the short and the long term, and a wide range of options is available. In particular, substantial economic potential exists for the mitigation of greenhouse gas emissions; mitigation may have net negative costs and many co-benefits (IPCC 2007; Rypdal et al. 2007; Williams 2007; Woodcock et al. 2007). However, adaptation and mitigation options also have associated implementation barriers, limitations and costs. One important consideration is that higher benefit-cost ratios could be achieved by introducing adaptation and mitigation measures at an early stage, rather than by attempting to retrofit infrastructure at a later date (IPCC 2007; Stern 2007). No single technology can provide all of the mitigation potential in any sector; however, the adoption of mitigation technologies in key sectors, including the transportation sector, is critical (Houghton 2009; Stern 2007).

Success in responding to the challenge of climate change depends upon co-operation and co-ordination between countries and the development of effective international climate change agreements (IPCC 2007; Yamin and Depledge 2004). Some progress has been made in the form of the United Nations Framework Convention on Climate Change (UNFCCC) and the Kyoto Protocol, and in the various negotiations to cover the post-Kyoto period from 2012, but the challenges of negotiating an adequate international climate change framework are far from being overcome (Adams 2009). Such a framework is necessary if emission trajectories for the stabilisation of greenhouse gas concentrations in the atmosphere are to be achieved; in turn, this requires steep, substantial cuts in national greenhouse gas emissions (Bows et al. 2006; Gössling and Upham 2009). The international community also faces the urgent challenge of creating a transparent, comparable carbon price signal worldwide, which is necessary for mitigation as it would generate incentives for producers and consumers to invest in low-carbon products, technologies and processes (Stern 2007). Whilst a wide range of policy instruments is available to governments to create incentives for mitigation, any given policy instrument has particular advantages and disadvantages, and various synergies and trade-offs exist between the adaptation and mitigation options (IPCC 2007). In general, the overall response to climate change will necessarily involve the use of a wide range of policy instruments to encourage changes in behaviours and lifestyles – including changes in consumption patterns – within the context of concerted, collective action at all levels of society (Houghton 2009; IPCC 2007; Stern 2007).

The Impact of Air Transport on Climate

Air transport affects climate in a variety of ways; those effects can be divided between the impacts of aircraft emissions, which represent the dominant issue, and a smaller but nonetheless significant group of effects due to the operation of airports and their ancillary services. The majority of the scientific studies of the impacts of aviation on climate have hitherto focused on aircraft emissions; many of those studies considered both subsonic and supersonic aircraft, although civil aviation is currently restricted to subsonic flight (Drake and Purvis 2001; IPCC 1999; Lee 2009). The effects of civil aircraft on climate can be divided into direct and indirect effects (Table 3.1). The direct effects include the release of substances into the atmosphere that cause radiative forcing: specifically, emissions of the greenhouse gas, CO_2, as an inevitable consequence of fossil fuel combustion, and emissions of aerosols, including soot and sulphate particles, which also have direct radiative forcing effects. In contrast, the indirect effects occur when aircraft emissions initiate other physical or chemical processes that in turn cause radiative forcing. Indirect effects include the enhancement of ozone (O_3) in the troposphere and the destruction of CH_4, both of which result from aircraft NO_x emissions. Other indirect effects are the formation of condensation trails (contrails) and the modification of cirrus cloud coverage (IPCC 1999). Thus, at cruise levels, aircraft have several main effects: (a) they alter the concentration of atmospheric greenhouse gases, including CO_2, tropospheric O_3 and CH_4; (b) they may trigger the formation of contrails; and (c) they may increase cirrus cloudiness. All of those effects contribute in different ways to climate change (IPCC 1999; Table 3.1). Whilst aircraft operations during the landing and take-off (LTO) cycle do not lead to contrail and cirrus cloud formation, they nevertheless generate emissions of greenhouse gases and particles; hence the LTO cycle should not be overlooked when considering the overall impact of aircraft on climate. An overview of the main pollutants emitted by aircraft has already been provided in Chapter 2; below, the effects of those emissions on climate are explained in greater detail.

Carbon dioxide (CO_2) is an important greenhouse gas covered by the Kyoto Protocol and it is the only long-lived greenhouse gas emitted in significant quantities by civil aircraft (UN 1998; Lee 2009; Figure 2.3). Aircraft emit CO_2 as a simple function of the quantity of fuel burned, with the result that the CO_2 emissions of individual aircraft operations are relatively easy to calculate (see Table 2.1). Increases in fuel consumption result in a linear increase in CO_2 emissions; conversely, improvements in aircraft fuel efficiency yield concomitant reductions in CO_2 emissions (Lee 2004). Whilst individual CO_2 molecules remain in the atmosphere for only around four years, the effective lifetime of anthropogenic CO_2 is much longer because it is not destroyed in the atmosphere but is redistributed between various carbon reservoirs: atmosphere, oceans (including oceanic biota), soil, terrestrial biota, rocks and ice. Exchanges of CO_2 between these reservoirs occur on widely varying timescales and are much longer than the atmospheric lifetime of individual CO_2 molecules (Houghton 2009). Overall, therefore, the atmospheric lifetime of CO_2 is extremely long and the gas becomes well-mixed and globally distributed in the atmosphere. Consequently, CO_2 emitted by aircraft – in common with all anthropogenic CO_2 – may interact with the climate system on the millennial timescale (Archer 2005; Houghton 2009; Lenton et al. 2006). Whilst CO_2 is not the strongest greenhouse gas, its relative abundance and long atmospheric lifetime make it the dominant component of the anthropogenic greenhouse effect and the most important greenhouse gas in international climate change policy debates (IPCC 2007; Stern 2007).

Table 3.1   The main effects of aircraft on climate

Effect	Significance
Emission of carbon dioxide (CO_2)	Carbon dioxide is the most important greenhouse gas emitted by human activities; it has a warming effect
Emission of nitrogen oxides (NO_x)	Nitrogen oxides cause the enhancement of tropospheric ozone (O_3), which is a greenhouse gas; NO_x also depletes methane (CH_4), another greenhouse gas
Emission of water vapour (H_2O)	The effects of water vapour emissions from civil aircraft are currently thought to be small, but they may be significant in the future – especially if a supersonic fleet is developed
Emission of soot particles	Soot particles absorb solar radiation and cause a small, localised warming of the atmosphere; they also act as cloud condensation nuclei in the formation of contrails and cirrus clouds
Sulphate particles derived from emission of sulphur oxides (SO_x)	Sulphate particles backscatter solar radiation, resulting in a small cooling effect; they also act as cloud condensation nuclei in the formation of contrails and cirrus clouds
Condensation trails (contrails)	Contrails are thought to have a warming effect on the atmosphere; they may evaporate rapidly or may persist and spread to form cirrus clouds
Enhanced cirrus cloudiness	Enhanced cirrus cloudiness has a potentially large warming effect on the atmosphere, but many scientific uncertainties remain about this issue

Aviation emissions of CO_2 currently amount to over 0.7Gt CO_2 annually, representing around 2–2.5 per cent of global anthropogenic greenhouse gas emissions (Lee 2009). Model results indicate that CO_2 emissions will have more than doubled by 2025 (Eyers et al. 2004; Macintosh and Wallace 2008); by 2050, under 'business as usual' projections, aviation CO_2 emissions are expected to represent 2.5 per cent of global greenhouse emissions (Stern 2007; WRI 2005; Yamin and Depledge 2004; Figure 2.4). However, given the recent commitments made by Kyoto-signatory states to reduce CO_2 emissions, projections indicate that, by 2050, aviation emissions could consume most or even all of national allowances, depending on the scenarios used, if international aviation emissions are taken into account (Bows et al. 2009; Owen and Lee 2006). Investigations undertaken during the EC TRADEOFF project indicated that the radiative forcing of global aviation CO_2 emissions in 2000 was $0.028W\ m^{-2}$ – and this is increasing rapidly (Lee 2004; Mayor and Tol 2010). Given the increases in aviation fuel consumption that are projected to accompany the sustained growth of air transport, CO_2 emissions from aviation are projected to increase dramatically by 2030, even with substantial technological improvements, and to increase by a factor of three over 1992 levels by 2050 (Horton 2006; Lee 2004; Figure 2.4). For this reason, Houghton (2009, 346) has stated that 'controlling the growing influence of aviation on the climate is probably the largest challenge to be solved in the overall mitigation of climate change.'

Nitrogen oxides (NO_x), which comprise nitric oxide (NO) and nitrogen dioxide (NO_2), are created by the oxidation of atmospheric nitrogen in the high temperature conditions occurring in the combustor of the aircraft engine, with some small quantities of NO_x arising from the nitrogen content of the fuel (Lee 2009; Rogers et al. 2002; see Chapter 2). Most of the NO_x emitted by aircraft is in the form of NO, but this is rapidly converted to NO_2 in the atmosphere; however, some NO_2 ('primary NO_2') is emitted directly by aircraft engines. The formation of NO_x during the combustion process is unavoidable, although NO_x emissions can be reduced by careful combustor design (Lee 2009; Rogers et al. 2002). Nevertheless, emissions of NO_x by aircraft have been forecast to increase by a factor of approximately 1.6 between 2002 and 2025 (Eyers et al. 2004). Emissions of NO_x from aircraft have two significant, indirect effects on climate: the enhancement of tropospheric ozone (O_3) and the reduction of methane (CH_4), both of which are greenhouse gases.

The catalytic production of tropospheric O_3 as a result of NO_x emissions occurs by several complex chemical processes (Brasseur et al. 1998; IPCC 1999; Lee 2004; 2009; Lee and Raper 2003; Rogers et al.

2002). The atmospheric residence time of NO_x depends upon the altitude of the emissions and is longer in the upper troposphere than at the Earth's surface; in turn, the longer residence time of NO_x at cruising altitudes allows more time for O_3 to be produced. Increases in O_3 concentrations in the upper troposphere also generate a stronger radiative forcing than do increases at lower altitudes (IPCC 1999). Furthermore, the tropospheric O_3 that is produced is less readily removed from the atmosphere than it would be by dry deposition at the Earth's surface (Lee and Raper 2003). Consequently, the production of tropospheric O_3 due to NO_x emissions is particularly efficient as a consequence of the levels at which aircraft typically cruise, and tropospheric O_3 may persist in the atmosphere for weeks (Lee 2004).

Besides the catalytic production of tropospheric O_3, aircraft emissions of NO_x result in the destruction of ambient CH_4: an effect that is significant because CH_4 is a potent greenhouse gas. Overall, therefore, aircraft emissions of NO_x produce a positive radiative forcing from the enhancement of O_3 and a negative forcing from the destruction of CH_4 (Lee and Raper 2003). Whilst those forcings are of comparable magnitudes, their effects on climate do not cancel because they have different geographical effects as a result of the different atmospheric lifetimes of the gases. Tropospheric O_3 has a relatively short atmospheric lifetime (of weeks to months) and does not become well-mixed globally; instead, tropospheric O_3 enhancement is concentrated in the Northern Hemisphere where more aircraft movements occur. In contrast, CH_4 is a long-lived, well-mixed greenhouse gas and the effect of its depletion by NO_x is global in scale. Some studies have suggested that the uneven distribution of tropospheric O_3 enhancement causes a stronger increase in surface temperature than if the O_3 enhancement were more uniformly distributed (IPCC 1999; Lee 2004; 2009; Stuber et al. 2001). Although the climate effects of tropospheric O_3 enhancement and CH_4 destruction due to NO_x emissions do not cancel, the reduction in CH_4 does partially offset the impact of another pollutant – CO_2 – since both of those greenhouse gases are long-lived and well-mixed in the atmosphere (Lee and Raper 2003). As a result of such uncertainties, Lee (2009) has emphasised that the effect of aviation NO_x emissions on climate remains open to debate.

Water vapour (H_2O) is also emitted by aircraft as a consequence of the combustion of kerosene. Water vapour is a powerful greenhouse gas and plays a critical role in the natural greenhouse effect of the Earth (Seinfeld and Pandis 2006). The troposphere is relatively humid due to natural hydrological processes, and the quantity of water vapour emitted by subsonic aircraft is extremely small compared with ambient concentrations, even if the larger projected emissions for 2025 or 2050 are considered (Eyers et al. 2004). Therefore, the climate effects of water vapour emissions from civil aircraft are currently small (Lee 2004; 2009; Lee and Raper 2003). However, the upper troposphere and lower stratosphere are drier, and the mid- and upper stratosphere are extremely dry; therefore, any introduction of water to the stratosphere is likely to cause atmospheric warming. Scenarios in which proposed fleets of supersonic aircraft operate at higher levels in the stratosphere could result in water vapour emissions from aircraft having a significant effect on climate (Lee 2004; 2009; Lee and Raper 2003).

Aerosols are suspensions of fine solid or liquid particles in a gas; they may be emitted directly as particles ('primary aerosol') or form in the atmosphere by conversion processes ('secondary aerosol'). Typically, atmospheric aerosols are considered to be those particles ranging from several nanometres to tens of micrometres in diameter (Seinfeld and Pandis 2006). The main aviation-derived aerosols are soot and sulphate particles, each of which has direct impacts on climate by the scattering and absorption of solar radiation (Seinfeld and Pandis 2006). Soot particles absorb incident solar radiation and cause local heating. Sulphate particles backscatter radiation, resulting in a cooling effect. However, both of those direct effects have been estimated to be relatively small (Lee 2004; 2009). In addition to their direct effects, particle emissions also result in indirect effects on climate through their role as cloud condensation nuclei in the formation of contrails and cirrus clouds (see below). Particle emissions are forecast to increase over the period 2002–25 due to the difficulty of sustaining significant technology improvements in this area (Eyers et al. 2004).

Contrails (condensation trails) are linear ice clouds that form behind aircraft due to the heating effect of aircraft engines, and due to the emission of water vapour and particles in the aircraft exhaust. Contrails form when the hot, humid exhaust gas from aircraft engines mixes with ambient cold, drier air, resulting in a sharp increase in relative humidity. Therefore, very cold (typically below $-40°C$) and humid (ice-supersaturated) ambient conditions are required for contrails to form (Lee 2009; RCEP 2002). Initially, water vapour emitted from the engine is deposited on particles in the exhaust, such as soot and sulphate particles. However, the contrail subsequently grows as ambient water vapour condenses onto the particles; thus most (around 98 per cent) of the ice constituting a persistent contrail originates from the ambient atmosphere (Lee 2009). The lifetime of a particular contrail is governed by the conditions of temperature and ice-supersaturation in the air masses through which an aircraft flies. Contrails may thus evaporate rapidly or may persist in the atmosphere and spread by diffusion and wind-shear, eventually forming cirrus clouds (see below). Therefore, contrails may persist in the atmosphere for periods ranging from seconds to hours (Lee 2004; Lee and Raper 2003).

The global coverage of linear contrails in 1992 was estimated to be about 0.1 per cent of the Earth's surface area, although contrail coverage may amount to 5 per cent regionally (Meyer et al. 2002; Sausen et al. 1998). The overall climate effect of contrails is the net result of both negative and positive radiative forcing effects; negative radiative forcing occurs as contrails reflect incoming solar radiation, whilst positive radiative forcing occurs as they also absorb infrared radiation from the Earth's surface. The precise balance of these two effects depends upon the optical properties of any given contrail. The IPCC (1999) reported that the radiative forcing due to linear contrails in 1992 was estimated to be 20mW m^{-2}, although subsequent studies have yielded lower estimates based on improved understanding of the optical properties of contrails. Nevertheless, significant uncertainties about the climate impacts of contrails remain (De Leon and Haigh 2007; Lee 2004; Lee and Raper 2003; Marquart and Mayer 2002; Minnis et al. 1999; 2004; Myhre and Stordal 2001). Contrail coverage is projected to increase more rapidly than the growth in aviation fuel consumption, for two reasons: (a) because air traffic will increase mainly in the upper troposphere, where contrails tend to form; and (b) because contrails may occur at higher temperatures – and hence more commonly – as a result of improvements in aircraft fuel efficiency (IPCC 1999; Lee 2009). Since the conditions necessary for contrail formation occur most frequently at night-time during winter, and because the majority of civil air traffic occurs in the Northern Hemisphere, the strongest effect of contrails on radiative forcing occurs at night-time during the Northern Hemisphere winter (December to February). Stuber et al. (2006) have demonstrated that, whilst night flights account for only 25 per cent of daily air traffic, they may contribute 60 to 80 per cent of the radiative forcing due to contrails. Furthermore, whilst winter flights account for only 22 per cent of annual air traffic, they may generate half of the annual mean radiative forcing due to contrails. Stuber et al. (2006) have argued that this disproportionate effect of night-time contrails during the Northern Hemisphere winter months on radiative forcing may be due to the fact that day-time contrails partially offset their overall warming effect by reflecting incoming solar radiation – an effect that disappears at night.

Cirrus clouds represent another, indirect climate impact of air transport. Aircraft may affect cirrus cloud coverage in two ways: (a) by the spread of persistent contrails by diffusion and wind-shear, leading to the formation of cirrus that is indistinguishable from natural cirrus (the primary effect); and (b) by the emission of particles that subsequently act as cloud condensation nuclei under more favourable atmospheric conditions of temperature and humidity, even a considerable time after the passage of the aircraft (the secondary effect). Of those two effects, the primary effect is more readily demonstrated (for instance, by the use of satellite imagery) and it can lead to extensive cloud coverage that would not have occurred without the aircraft emissions (Lee 2004; 2009; Lee and Raper 2003; Stordal et al. 2005). As with contrails, the net radiative effects of cirrus clouds depend upon the balance between their reflection of incoming solar radiation and their absorption of outgoing infrared radiation, with that balance depending in turn upon the optical properties of any given cloud. The IPCC (1999) has indicated that the effect of aircraft-derived cirrus cloud coverage is potentially large but also highly uncertain (see Lee 2009; Sausen et al. 2005). Nevertheless, a long-term increase in cirrus cloud coverage has been documented for which increasing levels of air traffic may have been partially responsible (Boucher 1999; Mannstein and Schumann 2005; Wylie et al. 2005). Various studies have demonstrated a significant correlation between air traffic and increasing cirrus cloud coverage; as much as an additional 1–3 per cent of cirrus cloud coverage per decade may be attributed to air traffic (Boucher 1999; Lee 2004; Stordal et al. 2005; Zerefos et al. 2003). Further research is required to clarify the climate effect of enhanced cirrus cloud coverage resulting from air transport.

By summing these various effects, it is possible to estimate the overall radiative forcing – and hence the overall climate response – due to air transport. The IPCC (1999) has published estimates of the radiative forcing due to aircraft emissions for 1992 and 2050; those estimates indicated that aviation was responsible for an overall radiative forcing of $0.05W \ m^{-2}$ in 1992 (excluding the effects of cirrus cloud enhancement, for which insufficient data were available), representing approximately 3.5 per cent of the overall radiative forcing of climate due to human activities at that time (ICAO 2001; IPCC 1999; Lee 2004). The IPCC (1999) reported that, by 2050, the radiative forcing due to aviation is projected to increase to $0.19W \ m^{-2}$, which would amount to around 5 per cent of all anthropogenic radiative forcing for a (moderate) reference emissions scenario (IPCC 1999; Stern 2007). For the full range of 2050 scenarios included in that study, the results indicated that radiative forcing due to aviation could range from 0.13 to $0.56W \ m^{-2}$, corresponding to between 2.6 and 11 times the radiative forcing due to aviation in 1992 (IPCC 1999). The projected large increase in the radiative forcing of climate due to aviation is primarily due to the sustained growth of the industry.

However, scientific understanding of climate change in general, and of the impacts of air transport on climate, is developing rapidly. Since the publication of the influential IPCC (1999) report, significant progress has been made in several areas. Assessments of the radiative forcing due to aviation, for instance, have been updated by the EC TRADEOFF project. Improved estimates of the radiative forcing due to tropospheric O_3 enhancement and CH_4 destruction (as a result of NO_x emissions) are now available; investigations undertaken as part of the TRADEOFF project indicated that both of those forcings were smaller than previously thought, a result that may be explained by improvements in the vertical and horizontal resolution of the models used (Lee 2004). Considerable attention has also been devoted to the task of clarifying the radiative forcing due to contrails and aviation-derived cirrus cloud coverage (Amanatidis and Friedl 2004; De Leon and Haigh 2007; Lee 2004; Stordal et al. 2005). The IPCC (2007) has reassessed the value of global anthropogenic radiative forcing, including radiative forcing due to aviation; for the year 2005, aviation radiation forcing was found to be 14 per cent greater (excluding the effect of aviation-induced cirrus cloud) than the figure reported for the year 2000 (Lee 2009; Lee et al. 2009). By the year 2050, the radiative forcing due to aviation could be three to four times greater than the 2000 value, with aviation then being responsible for 4–4.7 per cent of total anthropogenic radiative forcing (excluding the effect of aviation-induced cirrus cloud; Lee et al. 2009). Uncertainties remain in these estimates, however, and further research is in progress to determine the radiative forcing due to aviation with greater precision.

Whilst the overall radiative forcing due to air transport has now been estimated with greater accuracy than at the time of the publication of the authoritative IPCC (1999) report, such global estimates may still conceal important geographical effects. In the discussion of the climate impacts of NO_x emissions presented above, some geographical variations in the radiative forcing due to tropospheric O_3 enhancement and CH_4 destruction were highlighted. Other spatial variations – due to the greater efficiency of tropospheric O_3 enhancement at cruising altitudes than at the surface, and due to the greater incidence of contrail and cirrus cloud formation at higher latitudes – have also been acknowledged. Such effects may produce radiative forcing that is geographically concentrated in the Northern Hemisphere mid-latitudes (in regions that are characterised by more frequent aircraft operations) and close to the tropopause (at the typical cruising levels of civil aircraft). The IPCC (1999, 3) has drawn attention to the fact that, because 'some of aviation's key contributions to radiative forcing are located mainly in the northern mid-latitudes, the regional climate response may differ from that derived from a global mean radiative forcing.' However, although the impacts of aircraft on regional climate may be significant, those impacts have not yet been determined with sufficient accuracy (IPCC 1999).

Yet, despite these regional variations in radiative forcing, the impacts of air transport on climate are predominantly global in their scope, both because of the global nature of air transport operations and because the impacts of long-lived aircraft emissions can affect the entire climate system over extended time periods (Houghton 2009; IPCC 2007). Whilst the impact of air transport on climate change is global in its scope, though, the effects of climate change on human societies may be evident at regional and local scales as well as globally. Hence the geography of the impacts of air transport on climate is complex and requires further, detailed investigation. The effects of aircraft on the climate system vary according to their spatial scale (from the global effects of CO_2 emissions to the highly localised radiative forcing due to individual contrails), their lateral distribution (from the ubiquitous effect of CH_4 destruction to the to the regional concentration of tropospheric O_3 enhancement), their vertical distribution (from the well-mixed, long-lived greenhouse gases to the formation of cirrus clouds at cruising levels) and their timescale (from millennia in the case of CO_2 to seconds in the case of ephemeral contrails). The complexity of these effects makes attempts to estimate the net radiative forcing of climate due to aviation extremely challenging.

Scientists and policymakers have attempted to overcome such challenges through the use of a radiative forcing index (RFI). The use of an RFI was introduced by the IPCC in response to various difficulties that had been encountered in attempts to apply the concept of global warming potential (GWP) to aircraft emissions (IPCC 1995; 1999). An account of those problems is provided in IPCC (1999; see also Lee 2004); they included difficulties in attributing contrails to emissions of any particular greenhouse gas, variations in the radiative forcing due to NO_x emissions with location and season, fluctuations in the location and timing of short-lived gases and aerosols, and changes in the overall composition of the atmosphere with time. Whilst GWPs provide a useful means of comparing emissions of long-lived, well-mixed greenhouse gases such as CO_2 and CH_4, they are unsuitable for characterising the climate impacts of air transport which include other effects. In response to this difficulty, the RFI was developed; for a given activity, the RFI is defined as the ratio of the total radiative forcing to the radiative forcing due to CO_2 emissions alone

(IPCC 1999). Based on analysis of a range of emission scenarios for the period from 1992 to 2050, the IPCC (1999, 8–9) reported that the 'overall radiative forcing due to aircraft (excluding that from changes in cirrus clouds) [...] is a factor of 2 to 4 larger than the forcing by aircraft carbon dioxide emissions alone.' The RFI for aviation was reported to be 2.7 (range 1.9–4.0) in 1992 and it varied thereafter according to the scenario adopted (IPCC 1999; Lee 2004). As an approximation, the IPCC (1999, 213) stated that the RFI for aviation was 'about 3'. One important aspect of the RFI, therefore, is that it provided a means of expressing the climate impacts of aviation beyond those of CO_2 emissions alone.

The use of an RFI thus divides the climate impacts of aviation into those due to CO_2 emissions and those caused by a range of other, 'non-CO_2' effects. The latter group includes the impacts of NO_x emissions, water vapour, particles, contrails and aircraft-induced cirrus clouds. The distinction between the CO_2 and non-CO_2 effects of aviation on climate is important because considerable debate has focused on the scientific validity and policy implications of the use of an RFI. As Forster et al. (2006) have acknowledged, a notable example of the use of an RFI within air transport policy is the use of a factor of 2–4 (specifically 2.5) by the UK Government in its Aviation White Paper, *The Future of Air Transport*. In that document, the UK Government stated that the impact of aviation on climate is 'thought to be 2–4 times greater than that from CO_2 alone' (DfT 2003c, 40; *cf.* DfT 2008). Following the publication of early estimates of an RFI by the IPCC (1999), studies as part of the EC TRADEOFF project calculated the RFI for aviation to be approximately 1.9, although that figure may have been greater if the effects of aviation-induced cirrus cloud were included (Lee 2004; Sausen et al. 2005). As yet, some policy instruments – such as emissions trading schemes – that are intended to mitigate the climate impacts of air transport have focused only on the effects of aviation CO_2 emissions. Whilst CO_2 emissions represent the most significant anthropogenic impact on climate in general, the situation with respect to aviation is more complicated: CO_2 emissions form only a part of the total climate impact of the industry. However, whilst some attempts have been made to devise a robust and scientifically defensible method of evaluating (and mitigating) the non-CO_2 effects of air transport, some authors have argued that attempts to include the non-CO_2 effects of aviation in policy instruments such as emissions trading schemes are premature due to the fact that the magnitude of those effects is still not known with sufficient certainty (Forster et al. 2006).

The preceding discussion has focused on the climate impacts of aircraft emissions, which represent the majority of the climate impacts of air transport. However, other climate impacts occur due to the operation of airports and their ancillary services (Hooper et al. 2003). The construction, operation and maintenance of airports results in the emission of greenhouse gases and aerosols due to energy conversion and supply, the production and transport of materials for consumption within airports, the combustion of fossil fuels and the disturbance or destruction of vegetation. Additional greenhouse gases and aerosols are emitted by surface transport vehicles used on and around airport sites. Further emissions occur due to the provision of services at airports, including baggage handling, catering, cleaning, refuelling and waste management services. If surface transport vehicles are powered by diesel or petrol engines, they may make significant contributions to greenhouse gas and aerosol emissions at and around airports. Whilst such emissions and their climate impacts may be small in comparison with aircraft emissions, they are nonetheless significant – not least because they may represent a large share of the carbon footprint directly attributable to individual airports, and also because they may be relatively easy to mitigate. Overall, however, the impact of air transport on climate represents a formidable challenge to policymakers (Houghton 2009). Various options for responding to that challenge are considered in the next section.

Reducing the Impact of Air Transport on Climate

Policymakers face a major challenge in addressing the impact of air transport on climate. The air transport industry is international in its scope, is growing rapidly, is technologically mature and is characterised by long lead-in times and long in-service lifetimes. Whilst incremental improvements in fuel efficiency (with concomitant reductions in CO_2 emissions) may continue to accompany refinements in airframe and engine design, the potential to achieve more radical reductions of emissions is much more limited (Stern 2007). Yet, as Lee et al. (2009) have acknowledged, the introduction of radical technologies will be necessary if substantive reductions in aviation fuel use (and emissions) are to be achieved. The options for reducing the impact of air transport on climate are typically categorised in terms of technological, operational and policy options, although such distinctions are not clear-cut (IPCC 1999; Lee 2004). Technological approaches focus on achieving improvements in fuel efficiency through the use of improved airframe and engine designs, and through the use of alternative fuels. Operational approaches involve changing the way in which aircraft are flown, such as by selecting cruising levels to increase fuel efficiency or to avoid forming contrails; however, the introduction of revised operational procedures is not always supported by firm scientific evidence and must be subordinate to overriding safety considerations (Lee 2004). Given that technological and operational approaches are unlikely to offset the full impact of air transport on climate, at least on the decadal timescale, policy options to manage that impact are now being explored. Yet those policy options must negotiate complex issues of equity between nations and sectors, and they must provide effective environmental protection without jeopardising the important role played by air transport in promoting economic and social development – including access to international markets for producers in developing countries (Goldstein 2001; Raguraman 1995; UNCTAD 1999a; 1999b; 1999c). In this section, various technological, operational and policy options are considered in turn.

Engine and Airframe Technology Options

The main ways in which improved engine and airframe technologies could reduce the impacts of aircraft on climate are by: (a) improved fuel efficiency, which has the effect of reducing CO_2 emissions; (b) reducing NO_x emissions; and (c) reducing particle emissions. Technological advances have dramatically improved the environmental performance of aircraft engines and airframes; fuel efficiency has increased whilst the specific emissions (the emissions per passenger-kilometre) of most aircraft-derived pollutants have been reduced as a consequence (Bows et al. 2009; Thomas and Raper 2000). Subsonic aircraft manufactured in 1999 were approximately 70 per cent more fuel-efficient per passenger-kilometre than aircraft four decades earlier (ICAO 2007). Most of that improvement was achieved as a result of improvements in engine performance, with the remainder due to advances in airframe design. Incremental improvements in fuel efficiency are expected to continue to be made, with the IPCC (1999) stating that a projected 20 per cent improvement in fuel efficiency would occur by 2015, and a 40 per cent improvement by 2050, relative to aircraft manufactured in 1999. Internationally, major technology research programmes have been established with the aim of reducing emissions of NO_x during the LTO cycle dramatically, whilst simultaneously improving engine fuel efficiency (ACARE 2004; IPCC 1999).

Aircraft fuel efficiency can be increased through aerodynamic improvements, weight reductions and the use of more fuel-efficient engines. Airframe improvements aim to reduce aerodynamic drag, which has the side-benefit of reducing CO_2 emissions in addition to reducing operating costs (Lee 2004). Such improvements include advances in the use of laminar flow technology, riblets, wing-tip devices and high aspect-ratio/low sweep configurations; more radical airframe designs, such as the blended wing-body and the wing-in-ground-effect designs, are also being investigated (Air Travel – Greener By Design 2005; 2006; 2007; 2008; Bows et al. 2009; Viswanath 2002). Approaches based on weight reduction include the use of lightweight materials (carbon fibre or reinforced plastic structures) in the construction of aircraft components (Finley 2008; Marsh 2007). Other approaches focus on the use of adaptive structures in the airframe, which could allow aircraft to be operated consistently at their design conditions, leading to more fuel-efficient flight; such adaptive structures could potentially also reduce the complexity and weight of aircraft by dispensing with additional control surfaces (ACARE 2004). Overall, however, airframe modifications are unlikely to yield large reductions in aviation emissions (Lee 2004). Developments in engine technology appear to be more promising: research programmes are investigating several key areas: combustor sub-component performance, complex multi-staged combustors, single-stage combustor technology, fuel injection, enhanced fuel-air mixing, combustor liner coolant flow and hot-gas residence time (Lee 2004). However, whilst NO_x emissions could potentially be reduced through the use of 'lean-burn' combustion techniques (which could in turn reduce some of the indirect effects of aircraft on climate), such lean-burning, low emission combustors are susceptible to combustion instabilities (Lee 2009).

Reducing NO_x emissions using technological means is not straightforward, and it involves making trade-offs with the control of other pollutants, especially CO_2. Attempts to develop quieter and more fuel-efficient engines have prompted increases in the overall pressure ratio of engines. In general, higher operating temperatures and pressures lead to increased NO_x emissions, with the result that combustor technology must make commensurate progress in order to maintain or improve NO_x emission performance (Lee 2004). Improvements in the efficiency of engines may also lead to the increased production of contrails during cruise, and to cirrus cloud enhancement, due to the fact that more efficient engines induce the formation of contrails at higher temperatures (IPCC 1999; Lee 2009). Therefore, the design of future engines and airframes involves a complex decision-making process and many considerations must be balanced, including CO_2 emissions, NO_x emissions at ground level, NO_x emissions at altitude, water vapour emissions, contrail formation, cirrus cloud production and aircraft noise (IPCC 1999). Thus technological approaches to reducing the impact of air transport on climate involve balancing a range of aviation environmental impacts and making rational trade-offs – based on scientifically-robust data and on agreed priorities – between the management of those impacts. Overall, advances in airframe and engine technologies are likely to assist in reducing the impact of aircraft on climate; but, until radically new airframe or engine designs become available, the use of policy-based approaches will also be required (Peeters et al. 2009; RCEP 2002).

Fuel Options

Commercial civil aircraft require a high energy-density fuel, especially for long-haul flights, and conventional aircraft engines are designed to burn kerosene (Chapter 2). In 1999, the IPCC (1999, 10–11) stated that there 'would not appear to be any practical alternatives to kerosene-based fuels for commercial jet aircraft for the next several decades.' Aviation is thus highly dependent upon the availability of reliable sources of kerosene in the short- to medium-term, and there are unlikely to be radical changes in the fuel used by the global fleet of civil aircraft over the decadal timescale (RCEP 2002). Progress has been made in reducing the sulphur content of kerosene which, in turn, has reduced aviation emissions of SO_x and sulphate particles. However, reducing the sulphur content of kerosene also affects the lubricity of the fuel, so the extent to which fuel sulphur content can be reduced is limited (IPCC 1999). In the long-term, however, the air transport industry, in common with other transportation sectors, faces the problem of crude oil replacement during the twenty-first century and crude oil scarcity – or high oil prices – could potentially curtail the growth of the air transport industry by 2050 (Allen 1999). Therefore, considerable efforts are being made to develop new aviation fuels, and the potential to reduce the impacts of aircraft on climate through the use of alternative fuels to kerosene has received scrutiny (Bows et al. 2009; RCEP 2002). The availability of alternative fuels, such as liquid hydrogen (H_2), biofuels, synthetic fuels and liquefied natural gas (LNG), or the use of alternative power sources, such as fuel cells, could potentially reduce the radiative forcing due to aircraft, although many of those innovations would require new aircraft and airport infrastructure designs (ACARE 2004).

Biofuels have already been developed for aviation use. In February 2008, in an experimental trial, a Virgin Atlantic Boeing 747 aircraft was flown using a biofuel mixture of Brazilian babassu palm oil and coconut oil together with conventional kerosene (Upham et al. 2009). A vegetable-based kerosene, PROSENE, produced using a mix of soy, canola, castor, colza, sunflower and other oils has also been tested in Brazil (the use of hydrated alcohol as an aviation fuel has also been investigated; Simões and Schaeffer 2005). Research into the use of biofuels in aviation has also been undertaken by Boeing, Air New Zealand and Rolls-Royce. However, the issue has become complex and controversial because biofuel production is acknowledged to have other, negative implications for sustainable development, such as influences on food prices and availability, and it is unclear whether the production of biofuels has a greater impact on climate than the combustion of kerosene. Hence, if biofuels are to be suitable for aviation use (or in any other sector), they must not compromise food supplies or food prices, worldwide, and they must have the potential to reduce greenhouse gas emissions throughout their lifecycle. In response to those concerns, second-generation biofuels are being developed; those biofuels involve the growth of entire plants – such as the woody species used to produce Fischer-Tropsch kerosene – specifically for use as fuel, rather than the use of selected parts of food crops, and they may hold more promise for climate change mitigation. At present, however, biofuels are not yet regarded as a sufficient response to the challenge of climate change and they may still be incompatible with the principles of sustainable development (Marsh 2008; Mathiesen et al. 2008; Takeshita and Yamaji 2008; Upham et al. 2009; see Chapter 6).

More radically, the development of a fleet of aircraft fuelled by liquid H_2 has been proposed; experimental aircraft have been constructed and research programmes have examined the technological feasibility and potential environmental impacts of its use. The combustion of liquid H_2 emits no CO_2, very few particles and less NO_x than the use of conventional kerosene; however, the CO_2 emitted during the production of the liquid H_2 fuel must be accounted for in an overall environmental assessment of H_2 technology (Janić 2008; Lee 2009; Ponater et al. 2006; RCEP 2002). Furthermore, the use of liquid H_2 fuel by aircraft would increase emissions of water vapour and could result in the increased formation of contrails, and in greater radiative forcing due to contrails, than is currently the case. Further research into this issue is required (Gauss et al. 2003; Lee 2004; Marquart et al. 2003). Some research has been undertaken to investigate whether fuel additives could reduce contrail formation, but the results indicate that fuel additives are not a viable contrail mitigation option (Gierens 2007). Overall, improvements in fuel technology – like airframe and engine technology improvements – require substantial investment and are likely to be viable only in the long term; radical progress in reducing emissions using improved fuel technology appears to be unlikely in the short to medium term (RCEP 2002; Upham et al. 2009).

Operational Options

Operational options for mitigating the impact of aircraft on climate focus on operating aircraft in such a way as to minimise emissions (Bows et al. 2009; Drake 1974; Eyers et al. 2004; Simões and Schaeffer 2005). This involves configuring, loading, manoeuvring and maintaining aircraft in such a way as to maximise fuel efficiency: for instance, by reducing unnecessary aircraft weight, increasing load factors, ensuring high standards of aircraft maintenance, minimising route distances, optimising cruising speeds and levels, and minimising periods of holding (or 'stacking'; IPCC 1999). Accelerating the rate of fleet renewal is another method by which aircraft operators can improve their average fuel efficiency (Bows et al. 2009). Again, maximising fuel efficiency has the side-effect of reducing CO_2 emissions as well as the emissions of other pollutants. There are many reasons why aircraft may consume more fuel during flights than might strictly be required to complete each sector; some of those reasons relate to airline procedures, whilst others concern air traffic management (ATM) systems and procedures, including system-wide considerations such as the design of airspace. In general, system inefficiencies arise in the air transport system during all phases of flight – and during ground movements – as a result of congestion, leading to air and ground traffic constraints. The IPCC (1999) acknowledged that improvements in ATM and other operational systems and procedures could reduce aviation fuel consumption by between 8 and 16 per cent, with the large majority (6–12 per cent) of those reductions coming from ATM improvements that are anticipated to be implemented fully by 2020. Improvements in the efficiency of the ATM system will in principle mean that specific aircraft emissions are reduced, although the rate at which improved ATM systems and procedures are introduced depends upon the creation of the necessary institutional frameworks at an international level, such as the implementation of the Single European Sky initiative (EUROCONTROL 2007; IPCC 1999).

Various improvements to the ATM system have been proposed or developed. One ATM system improvement that has already been implemented is the reduction of the vertical separation of air traffic in European airspace and in the North Atlantic Flight Corridor (NAFC) from 2,000 feet to 1,000 feet. That innovation has had several consequences: the capacity of the airspace has been increased and, in principle, aircraft may fly closer to their most fuel-efficient cruising levels. However, by increasing the vertical concentration of air traffic, the coverage of contrails may have increased as more aircraft are now likely to fly in contrail-forming conditions (Lee 2004). More radical operational options focus on diverting aircraft away from 'environmentally sensitive' regions of the atmosphere: for instance, avoiding ice-supersaturated areas – which may be less than one flight level in thickness – in order to avoid contrail formation (Gierens et al. 2008; Lee 2004; 2009). The effects of relatively simple changes in operational practices have been investigated in the EC TRADEOFF project. For example, the complex effects of flying at higher or lower altitudes on contrail formation, on CO_2 and NO_x emissions, and on O_3 enhancement, have been estimated by various authors (Fichter et al. 2005; Grewe et al. 2002; Lee and Raper 2003; Sausen et al. 1998). Flying at lower speeds also offers a potential means of achieving emissions reductions, although this involves compromising journey times and customer service levels (ACARE 2004; Drake 1974). A more realistic option may be the elimination of periods of holding, a feature of congested airspace in which aircraft are directed into holding patterns ('stacks') during approach until a landing slot becomes available. Holding is usually a temporary effect affecting only the busiest airports at certain times of the day; nevertheless, British Airways reported that the additional fuel burned during holding around three major UK airports over a one-year period from 2000–2001 amounted to 1.2 per cent of total fuel consumption by the airline (Eyers et al. 2004).

Another operational means of reducing specific emissions from aircraft is to increase the load factor of a flight, by eliminating non-essential weight and by maximising the payload (IPCC 1999). In this way, maximum utility can be obtained for a given level of environmental impact. Loading aircraft efficiently involves a combination of (a) minimising the weight of the aircraft before its payload is stowed (for instance, by minimising the carriage of unusable fuel) and (b) maximising the passenger seat occupancy or the cargo load. Efficient loading of aircraft can also be increased by avoiding tankering – the practice of enplaning more fuel than a particular flight requires so that it can be used for subsequent flights (Drake 1974; Eyers et al. 2004). The effect of tankering is to increase aircraft mass and thus fuel consumption and emissions. Tankering occurs for commercial reasons (perhaps because of substantially higher fuel costs at the destination airport which offset the cost of the increased fuel burn), although there may also be operational reasons for carrying surplus fuel (such as the desire to minimise turnaround time, to make an allocated runway slot, or to avoid refuelling in places where fuel availability or quality is not assured). Accurate data about the extent of fuel tankering are not readily available, as data about aircraft operating mass are commercially sensitive and notoriously difficult to obtain. Nevertheless, British Airways estimated that additional fuel burn due to tankering is around 0.5 per cent of total aircraft fuel consumption, although this depends on the aircraft type, on the flight distance, and on how many tankering flights are flown in sequence (Eyers et al. 2004; IPCC 1999).

Many other operational improvements could also contribute to emissions reductions. Improved aircraft maintenance could ensure that high levels of fuel efficiency are maintained throughout the service life of the airframe and engines (Curran 2006). Reducing the time that aircraft spend taxiing could yield fuel efficiency improvements in the range 2–6 per cent (IPCC 1999; Marín 2006). Many initiatives relating to ATM systems and procedures could potentially reduce emissions: improved communications, navigation and surveillance and air traffic management (CNS/ATM) systems; arrival management (AMAN) and departure management (DMAN) systems; collaborative decision making (CDM) systems; continuous descent approaches (CDAs) and low power, low drag (LP/LD) approaches, which may have the effect of minimising emissions during descent; and expedited climb departure procedures, which are designed to allow aircraft to climb rapidly to their optimal cruising levels without encountering climb restrictions (Auerbach and Koch 2007; Dobbie and Eran-Tasker 2001; ICAO 2004). Other operational efficiencies could potentially be achieved by the use of hub-bypass route planning (Morrell and Lu 2006) and by the use of fixed electrical ground power (FEGP) in preference to kerosene-burning auxiliary power units (APUs). Overall, however, operational options to reduce the climate impacts of aviation are generally in their infancy and more research is needed into their effects (Lee 2009; Lee and Raper 2003). Improved operational efficiency may even result in airports and routes attracting additional air traffic, thereby negating any emissions reductions that might otherwise have been gained (IPCC 1999). Consequently, the DTI (1996, 10) has acknowledged that 'operational measures will not offset the impact of the forecast growth in air travel'; whilst operational options may form part of an overall strategy to mitigate the impacts of air transport on climate, they are not a sufficient response to that challenge.

Policy Options

A wide range of regulatory, market-based and voluntary measures designed to reduce the impacts of air transport on climate can be grouped together within the category of policy options (Bishop and Grayling 2003; Cairns and Newson 2006; Daley and Preston 2009; Hewett and Foley 2000; Roberts 2004). It is clear that improvements in aircraft and engine technology, fuel technology and operational systems and procedures – including the increased efficiency of ATM systems and procedures – could mitigate some of the impacts of aviation on climate, but that those advances will not fully offset the emissions resulting from the projected growth in aviation (IPCC 1999). Given the high abatement costs of the sector and the limited potential for radical technological or operational solutions to be found in the short to medium term, success in reducing the impact of air transport on climate therefore depends upon the formulation and implementation of effective policy (Bishop and Grayling 2003; Bows et al. 2009; Daley and Preston 2009). Many instruments are available to policymakers, some of which have received scrutiny from a wide range of stakeholders. Specifically, the IPCC (1999) has highlighted the following policy options to reduce emissions: (a) more stringent aircraft engine emissions regulations; (b) removal of subsidies and incentives that have negative environmental consequences; (c) market-based options such as environmental levies (charges and taxes) and emissions trading; (d) voluntary agreements; (e) research programmes; and (f) substitution of aviation by other transport modes. However, not all of those options have been fully investigated or proven in relation to aviation (IPCC 1999). Proposals to cap aviation emissions, to impose taxes and emissions charges, to use or remove subsidies, to issue tradable permits for aviation emissions and to encourage the use of voluntary agreements have been made by various commentators (Bishop and Grayling 2003; IPCC 1999; Pastowski 2003). Such proposals have individual strengths but they are also problematic for a variety of reasons. Regulatory approaches face the problem that air transport is an international industry that spans national jurisdictions – and nations have varying capacities to monitor and to enforce environmental standards. Market-based approaches must negotiate difficult issues related to the varying competitiveness of air transport service providers, including the need to internalise differing costs of pollution whilst facilitating access to international air transport markets on an equitable basis between nations. Voluntary approaches – such as the use of carbon offsetting schemes and voluntary codes of conduct – face the criticism that they are too weak to catalyse the profound behavioural change that is required to ensure that air transport is compatible with the requirements of sustainable development (Broderick 2009; Hewett and Foley 2000). Various policy options, together with some of the issues associated with each, are discussed below.

Regulatory approaches   Regulatory approaches typically involve the imposition of standards, methods of enforcement and sanctions; such approaches are frequently used in environmental management, especially where pollutants pose a risk to human health. Aviation emissions are subject to some regulatory standards as part of the ICAO engine certification process. Aircraft emissions of HCs, CO, NO_x and smoke during the LTO cycle may not exceed specified values (as measured using relatively small samples of newly-manufactured engines); those certification limits provide a starting point for the management of aviation emissions (IPCC 1999). The certification standards are negotiated and agreed through the ICAO Committee on Aviation Environmental Protection (CAEP). The first CAEP emissions standards (the CAEP 1 regulations) were agreed in 1981 and took effect in 1986 and subsequent standards have increased the levels of stringency applied to new engines, with the result that current NO_x standards are approximately 40 per cent more stringent than the original standard (ICAO 2007; Lee 2004). Currently, of the pollutants regulated by ICAO, only the reduction of NO_x emissions represents a significant technological and environmental challenge (Lee 2004). (The ICAO-regulated pollutants do not include CO_2 – a point that is discussed below.) Although the reduction of these emissions is targeted at the LTO cycle rather than at typical cruising levels, technological improvements that reduce NO_x emissions over the LTO cycle tend also to reduce specific emissions during the cruise (Lee 2004). However, Lee (2004) has pointed out that the regulatory parameter for NO_x varies with engine overall pressure ratio, with the result that, for the global fleet, the emission factor for NO_x increased from 1992 to 1999 and is forecast to continue increasing until at least 2015 (Eyers et al. 2004; Lee 2004; see Chapter 2). Reducing aircraft NO_x emissions has been hindered by innovations designed to create quieter and more fuel-efficient engines, which have in turn involved increasing the overall pressure ratio of engines. New aircraft engines are now required to meet relatively stringent NO_x emissions standards, but older aircraft tend to remain in service long after the introduction of new regulations with the result that, over the entire global fleet, NO_x emissions continue to increase – as do their indirect effects on climate.

The ICAO engine certification standards therefore influence aviation impacts on climate via their effects on NO_x emissions, but they do not regulate emissions of the most significant pollutant from the point of view of radiative forcing: CO_2. The UN Framework Convention on Climate Change (UNFCCC) and its Kyoto Protocol are presently the dominant mechanism for managing the impact of anthropogenic CO_2 emissions (UN 1992; 1998). In the Kyoto Protocol, however, a distinction is made between domestic and international bunker fuels. Domestic aviation CO_2 emissions are counted as 'transport emissions' and are included in national inventory totals, whilst emissions from international aviation are reported separately and do not count towards national greenhouse gas totals under the UNFCCC. At present, the separate reporting of CO_2 emissions from international aviation bunker fuels in the annual national inventories of Annex I Parties is mandatory under the UNFCCC and international aviation emissions 'should' be reported by Non-Annex I Parties 'to the extent possible, and if disaggregated data are available', but those emissions are excluded from Annex I Parties' quantified commitments (UN 1992; 1998; Yamin and Depledge 2004; see also Lee 2004; Lee and Raper 2003). Given that approximately 60 per cent of global civil aviation emissions in 1992 resulted from international aviation, such an omission is highly significant (Lee and Raper 2003). Nonetheless, the Kyoto Protocol makes some provision for the inclusion of international aviation CO_2 emissions, stating in its Article 2: 'The Parties included in Annex I shall pursue limitation or reduction of emissions of greenhouse gases not controlled by the Montreal Protocol from aviation and marine bunker fuels, working through the International Civil Aviation Organization and the International Maritime Organization, respectively.' (UN 1998, 2)

Thus the UNFCCC determined that emissions from international bunker fuels shall be regulated through ICAO rather than directly within the Kyoto Protocol (Yamin and Depledge 2004). Subsequently, the UNFCCC, through its Subsidiary Body on Scientific and Technical Advice (SBSTA), requested that ICAO's CAEP investigate potential methods by which international aviation could be brought within the Kyoto Protocol (Gander and Helme 1999; Lee and Raper 2003).

The responsibility of ICAO to regulate aviation emissions is laid out in the ICAO Assembly Resolution on Environmental Protection (Resolution A33–7), particularly in Appendices H and I:

> Resolution A33–7 mandates the ICAO Council to continue to study policy options to reduce or limit the environmental impact of aircraft emissions, placing special emphasis on technical solutions and consideration of market-based measures, taking into account potential implications for developed and developing countries. (Yamin and Depledge 2004, 86–87)

That task falls within the remit of CAEP. However, in January 2001, CAEP decided not to impose an ICAO standard to limit aircraft CO_2 emissions, citing the diversity of aviation operations and the fact that (in principle) market pressures already ensure that aircraft are fuel-efficient. Instead, ICAO has encouraged contracting states to use voluntary measures to mitigate the climate change impacts of aviation;

ICAO has pledged to facilitate this process by developing appropriate guidelines (Yamin and Depledge 2004). Hence, in relation to the issue of climate change, international aviation currently remains exempted from any fixed limits or caps of its greenhouse gas emissions under the Kyoto Protocol. As a result, limited progress has been made in managing the greenhouse gas emissions of the sector since the Kyoto Protocol was signed (Faber et al. 2007). Discussions are ongoing to determine the potential for aviation to be subject to emissions limits in an international post-2012 climate agreement, although it is unclear whether such an approach will be politically acceptable given the importance of growing demand for international air transport for the economic development of nations (see *The Economist*, 10 June 2006, 10). The regulation of CO_2 emissions arising from international aviation raises three main issues: (a) the difficulty of producing adequate and consistent emission inventories; (b) the difficulty of allocating emissions to countries; and (c) the difficulty in devising suitable policy measures to control emissions (Yamin and Depledge 2004). To this list could be added a further, complicating issue: the fact that aircraft have other effects on climate besides those of their CO_2 emissions alone.

One key task for policymakers, therefore, is to determine how emissions from international aviation bunker fuels can be included in national greenhouse gas inventories. The allocation of those emissions is not a straightforward matter and various allocation methodologies have been proposed. In 1996, the UNFCCC identified eight options for the allocation and control of greenhouse gas emissions from international bunker fuels, including those used in aviation:

- Option 1 – No allocation;
- Option 2 – Allocation of global bunker fuel sales and associated emissions to countries in proportion to their national emissions;
- Option 3 – Allocation according to the country where the bunker fuel is sold;
- Option 4 – Allocation according to the nationality of the transporting country, or to the country where the aircraft or ship is registered, or to the country of the operator;
- Option 5 – Allocation according to the country of destination or departure of an aircraft or vessel; alternatively, emissions related to the journey of an aircraft or vessel could be shared by the country of departure and the country of arrival;

- Option 6 – Allocation according to the country of departure or destination of passengers or cargo; alternatively, emissions ,related to the journey of passengers or cargo could be shared by the country of departure and the country of arrival;
- Option 7 – Allocation according to the country of origin of passengers or of the owner of cargo;
- Option 8 – Allocation to a country of all emissions generated in its national space.

In the first option, international emissions are not allocated to individual countries but remain in the international sphere. In the other seven options, various criteria are employed for allocating international emissions to individual Annex I countries. Of the eight proposals listed above, Options 2, 7 and 8 were subsequently considered by the UNFCCC to be less attractive options on the basis of equity, tractability and efficiency, respectively (Lee et al. 2005; Owen and Lee 2005; 2006). The four basic allocation options that have being explored further are: (a) no allocation to national inventories; (b) allocation according to the county where the fuel is sold; (c) allocation according to the nationality of the airline/aircraft operator or aircraft registration; and (d) allocation according to the country of departure or destination of the aircraft (Yamin and Depledge 2004). Overall, however, progress with respect to the allocation options has been hindered by a lack of consensus among states, and further work is being undertaken to improve the transparency, accuracy and comparability of calculations of international bunker fuel emissions (Yamin and Depledge 2004).

Market-based approaches   Market-based approaches are based on the principle of creating economic incentives and disincentives for particular activities. With such approaches, polluters are not prohibited from causing environmental damage but they incur financial penalties for doing so and thus are encouraged to bring environmental costs within the scope of their decision-making. At the same time, market-based approaches may be used to make 'environmentally desirable' courses of action more advantageous to polluters. Market-based policy instruments include a range of incentives and disincentives; the main types used in environmental management are taxes, charges, subsidies and tradable (or marketable) permits. All of these types of policy instrument are either already in use or under consideration as a response to aviation environmental impacts, and some will almost certainly form part of future policy frameworks to manage the impacts of air transport on climate. Yet implementing market-based mechanisms is not easy:

complex issues must be negotiated, such as the differing competitiveness of air transport service providers and the need to internalise varying costs of pollution whilst facilitating access to international air transport markets on an equitable basis between nations.

Market-based approaches are based on the concept of a cost of carbon which ideally reflects the full cost of the environmental damage caused by CO_2 and other emissions (Stern 2007). The introduction of an appropriate cost of carbon could provide a strong incentive within the sector to innovate in the areas of airframe and engine efficiency, as well as operational systems and procedures, in order to reduce emissions. However, since air transport is international in its nature, the selection of an appropriate carbon pricing instrument is not straightforward (Bishop and Grayling 2003; Hewett and Foley 2000; Stern 2007). Nevertheless, as Stern (2007, 549) has acknowledged; 'extending the coverage of carbon pricing and other measures to international aviation will become increasingly important.' Horton (2006) investigated the growth of CO_2 emissions from civil aircraft to the year 2030 under a range of different scenarios; his study demonstrated that different costs of CO_2 could influence the growth of emissions, with the implication that the level at which a carbon tax is set could be an important consideration in any attempts to reduce emissions. Horton (2006) analysed five scenarios ranging from no fuel efficiency improvements to a realistic technology scenario plus a $100 per tonne cost of CO_2. Overall, in that study, the total annual distance covered by the global civil aircraft fleet was projected to grow by 149 per cent from 2002 to 2030, with seat-kilometres forecast to grow by 229 per cent. Under the scenario with the most rapid rate of technological advance (that with $100 per tonne cost of CO_2), emissions of CO_2 were projected to be 22 per cent less in 2030 than under the scenario without the extra incentives to develop new technologies. However, even in that relatively optimistic case, the CO_2 emissions in 2030 were projected to be almost double those in 2002 (Horton 2006).

One possible approach to tackling the impacts of air transport on climate change could involve setting (high) carbon taxes on aviation. The idea of fuel taxation has prompted fierce debate within the aviation industry, given the sensitivity of air transport to kerosene prices. However, particular difficulties exist in attempting to co-ordinate an international tax on aviation fuel. Article 24 of the Chicago Convention prohibits the taxation of fuel used for international flights; fuel taxation is also precluded by the many existing bilateral air service agreements (ASAs) that regulate international air services between nations (Stern 2007). Whilst individual nations are permitted to implement their own environmental fuel charges, few have done so for domestic aviation and the fuel used in international aviation remains untaxed (Cairns and Newson 2006; Carlsson and Hammar 2002; ICAO 2006b; IPCC 1999; Mendes and Santos 2008; Pearce and Pearce 2000; Seidel and Rossell 2001; Wit et al. 2004). Nonetheless, international co-ordination of an aviation fuel tax would be essential to avoid economic distortions and 'carbon leakage', by which taxes could induce travellers to switch to different carriers, could prompt air carriers to change their routes, or could encourage practices such as tankering (carrying excess fuel during flights in order to avoid refuelling at airports where fuel taxes are levied). Such economic distortions could have the net effect of increasing aviation emissions, thus negating the original purpose of the tax (Wit et al. 2004).

Other forms of taxation could be used to restrain demand for air travel and thereby curb emissions: for instance, imposing Value Added Tax (VAT) on air international tickets, which are currently VAT-free (although this option is regarded as logistically complex) or through increases to the Air Passenger Duty (APD) (an option that is regarded as a 'blunt instrument'; Cairns and Newson 2006; DfT 2003c). The level of any tax is critical; Cairns and Newson (2006, 53) acknowledged that 'very large increases in fares would be needed to make a difference to demand' and that such increases would be politically unacceptable. Mayor and Tol (2007) have demonstrated that the recent doubling of the APD in the UK has actually increased, rather than decreased, CO_2 emissions because it reduced the relative price difference between near and far holidays; however, Mayor and Tol (2007) argued that CO_2 emissions would fall rather than rise if the same revenue was raised using a carbon tax rather than a boarding tax. For the reasons mentioned above, levels of taxation in the aviation sector globally are currently low in comparison with road transport fuel taxes. In turn, those low relative taxation levels contribute to congestion and to capacity limits at airports, which Stern (2007) has argued represents a form of rationing – which is an inefficient way of regulating demand. The UK Government has acknowledged that 'the global exemption of aviation kerosene from fuel tax is anomalous, but a unilateral approach to aviation fuel tax would not be effective in the light of international legal constraints' (DfT 2003c). Emissions charges offer an alternative to fuel taxation and represent a straightforward means of increasing the cost of environmentally destructive practices. Emissions charges face fewer legal obstacles than fuel taxes because they are not explicitly precluded by legally binding agreements; furthermore, if emissions charges were introduced on an en route basis, there would be a smaller likelihood that tankering would occur in response (Hewett and Foley 2000; IPCC

1999, 346; Wit et al. 2004). Alternatively, whilst taxes or charges, such as landing charges, are blunt instruments for cutting CO_2 emissions, they could be differentiated – for example, according to distance flown – in order to improve their effectiveness (Stern 2007).

Subsidies are another type of market-based instrument that are designed to provide direct incentives for environmental protection. Subsidies have been widely used in attempts to control pollution and to mitigate the financial impacts of regulations by helping polluters to meet the costs of compliance; they may take the form of grants, loans or tax allowances. In aviation, subsidies could be used to accelerate fleet replacement or to promote the development and use of alternative fuels – as well as other technologies – to reduce the environmental impacts of aircraft (Lambert 2008; RCEP 2002). However, aviation already benefits from a range of economic incentives that have allowed the industry to avoid paying the full environmental costs of its activities. Many parts of the air transport industry are subsidised and receive tax exemptions: for instance, jet fuel for international flights has historically been exempted from taxation; international air tickets are exempted from VAT; airlines and new regional airports receive direct aid; the industry receives investment grants, government loans, infrastructure improvement subsidies and launch aid; aircraft landing fees are cross-subsidised with parking and retail revenues at airports; and the manufacture of aircraft is exempted from VAT (EC 2006; Peeters et al. 2006; T&E 2006). Several countries levy ticket or fuel taxes on domestic flights but those measures do not compensate for the general tax exemption of the sector (T&E 2006). In general, the trend in environmental policy for aviation should ideally be towards the removal of current subsidies and privileges within the sector rather than the creation of new ones (Peeters et al. 2006).

Tradable (or marketable) permit schemes represent another market-based policy instrument; those schemes provide polluters with incentives to reduce pollution by creating new markets with defined property rights (Gander and Helme 1999; Seidel and Rossell 2001). Tradable permit schemes operate on a simple principle: (a) a total level of pollution is defined for a specific region; (b) permits equalling that level are distributed amongst polluters in the region; and (c) those permits are then traded, either amongst polluters or between the operational sites of individual polluters. Hence tradable property rights to pollute the environment are assigned to polluters. The overall level of emissions for the industry is fixed (as with a regulatory standard); but, once the market is operating, the distribution of permits – and thus of emissions – is determined by the polluters trading in the market. The trade in permits should in principle result in a concentration of emission reductions at those sources where they can be achieved at least cost. Polluters faced with high abatement costs may purchase permits from polluters who have achieved emission reductions, as that course of action is cheaper than incurring abatement costs. For aviation, which would incur high abatement costs for its impacts on climate, the possibility of buying additional emission permits from other sectors could offer a way of continuing to operate – and even to grow – despite increasing constraints on emissions at national or international levels (assuming that sufficient emission reductions can be achieved by other sectors; DfT 2004; Lee 2004; Upham and Gössling 2009).

Under the Kyoto Protocol, the use of tradable permits within emissions trading schemes is evolving as an important element of international climate policy. The largest scheme in the world is the EU Emissions Trading Scheme (ETS), currently in its second trading period. Emissions trading schemes represent one preferred route by which international aviation CO_2 emissions could be brought under the Kyoto Protocol; the ICAO Assembly has endorsed the development of an open emissions trading system for international aviation (DfT 2003c; Gander and Helme 1999; Upham and Gössling 2009). Aviation was not included in the first round of the EU ETS, but, in December 2006, the EC adopted a proposal to include aviation within the EU ETS. That proposal aims to bring aviation into the trading scheme in two stages, commencing in 2011 with intra-EU flights (domestic and international flights between EU airports) and then expanding in 2012 to include all international flights arriving or departing from EU airports (CEC 2006). A range of logistical issues remains to be resolved, particularly in relation to trade rights, the initial allocation of permits, the avoidance of 'windfall' benefits due to the over-allocation of permits, the coverage of the scheme, and the possible use of a factor to account for the non-CO_2 climate effects of aviation (DfT 2004; Forster et al. 2006; Karmali and Harris 2004; Lee 2004; Lee and Sausen 2000; IPCC 1999; Mendes and Santos 2008; Wit et al. 2005). The last of those issues is particularly significant: emissions trading schemes incorporating only CO_2 emissions do not account for the additional, non-CO_2 effects of aviation on climate, with the result that the purchase of permits by the aviation industry could increase, rather than reduce, total radiative forcing (Lee and Raper 2003). To account for the non-CO_2 effects of aviation within an emissions trading scheme, a factor (such as the radiative forcing index) could be used, although this remains contentious due to scientific uncertainties about the robustness and accuracy of the radiative forcing index for aviation (Forster et al. 2006; Lee 2004).

Overall, market-based approaches probably represent the most promising policy instruments for reducing the impacts of air transport on climate. In particular, emissions trading schemes have been described as the most advanced global policy instrument to tackle climate change (Hewett and Foley 2000). However, individual market-based approaches are likely to be most effective if used in conjunction with other measures; Stern (2007) has acknowledged that emissions trading could be used in tandem with taxation or with additional complementary measures such as voluntary co-operation and the sharing of best practice in emission reduction. Yet whichever instrument is chosen, the implementation of appropriate market-based policy 'is likely to be driven as much by political viability as by the economics' (Stern 2007, 388). Important areas in which further work is ongoing include the development of guidance for countries wishing to develop emissions trading with respect to aviation, and the development of a better understanding of the potential trade-offs involved in managing CO_2 emissions and other environmental impacts. However, Stern (2007) has also acknowledged that market-based measures do not, of themselves, regulate emissions. Further, careful investigation is required to determine the most effective way in which market-based – and other – approaches could be used to reduce the radiative forcing due to aviation.

Voluntary approaches Voluntary approaches to mitigating the impacts of air transport on climate also represent an important group of policy instruments (Stern 2007). Policy approaches based on voluntary measures rely upon organisations and individuals making decisions that take account of environmental concerns, even in the absence of direct regulatory requirements or economic incentives. Such voluntary decisions may be motivated by a variety of concerns. Polluters may believe that working co-operatively with regulators is more likely to lead to sympathetic regulation of their operations, and polluters may perceive greater opportunities to influence the regulatory process if they can demonstrate substantial voluntary efforts to improve their environmental performance. Polluters may adopt voluntary measures in anticipation of stricter regulation in the future, especially if early adoption offers them a competitive advantage; they may voluntarily adopt cleaner processes in order to standardise their operations across countries or regions; and they may seek to maximise their access to worldwide markets by adopting processes that would comply with the environmental regulations of the strictest country. Ultimately, organisations may voluntarily improve their environmental performance if they believe that consumer expectations require such action. Hence companies can improve their consumer relations and brand images by demonstrating corporate responsibility, either environmentally or socially (Broderick 2009). In relation to aviation and climate policy, voluntary measures currently focus on the use of carbon offsetting, on commitments to achieve 'carbon neutrality' and on the adoption of a range of broader corporate responsibility initiatives – including 'eco-labelling' initiatives.

Carbon offsetting has become a widespread response to the challenge of climate change. In 2006, an estimated 1.5 million people in the UK paid to offset the emissions of a flight (*New Scientist*, 24 February 2007, 35; see also Bayon et al. 2007; Broderick 2009; Gössling et al. 2007; Jardine 2005; Rousse 2008). Many issues are associated with carbon offsetting, especially in relation to aviation; those issues relate mainly to the measurement of emissions and to the permanence and credibility of offsets. The main areas of concern are that offsetting is not a sufficient measure to address climate change, for many reasons: it does not address all of the climate impacts of aviation; it requires accurate measures of the emissions generated and those saved elsewhere; it requires an appropriate price to be put on one tonne of CO_2e (CO_2e refers to 'CO_2 equivalent' – the quantity of a greenhouse gas that has the same radiative effect as one tonne of CO_2); it requires demonstrating additionality, which represents a considerable challenge; offsetting schemes are unregulated, may be overpriced and are vulnerable to fraud; the schemes can be inefficient; offsets may not be permanent; the schemes may create problems of leakage; the projects may have mixed sustainable development side-effects; and the schemes may be a distraction from the real challenge of reducing emissions, and so could delay the transition to a low-carbon economy. Given those issues, offsetting is now acknowledged to be a highly problematic response to the challenge of climate change (Broderick 2009; Brouwer et al. 2007; Daley and Preston 2009; DEFRA 2007a; Friends of the Earth et al. 2006; *The Guardian*, 16 June 2007, 15).

Nevertheless, the use of carbon offsetting increasingly forms an element of corporate commitments to achieve 'carbon neutrality', however that may be defined. Within the aviation industry, such commitments are now being made by some airport operators; yet those commitments involve subtle issues of definition and coverage. Airport operators define their spheres of responsibility and influence in various ways, and in particular they differ in their 'ownership' of aircraft emissions in the vicinity of the airport. Some airport operators accept responsibility for aviation emissions produced throughout the LTO cycle, whilst others restrict their responsibility to those emissions generated while aircraft are parked at the gate or are manoeuvring on the apron and taxiways. Such differences in coverage have significant implications for the magnitude of the carbon burden to be mitigated, and for the possibility that some emissions may not be apportioned to any particular polluter. However a given airport's sphere of responsibility is defined, airport operators have a much larger sphere of influence over airlines and, in addition to demonstrating carbon neutrality for their own operations, could focus greater efforts on encouraging airlines to achieve emission reductions – for example, by accelerating their fleet renewal processes.

Summary

Air transport currently has a small but nonetheless significant – and increasing – impact on climate. In addition to the effects of airports and their ancillary services, aircraft affect climate in several ways. Emissions of CO_2 have direct radiative effects on climate. Small, direct radiative effects also occur due to emissions of soot and sulphate particles. In addition, aircraft emissions have a range of indirect effects: the enhancement of O_3 in the upper troposphere and lower stratosphere, and the destruction of ambient CH_4, as a result of NO_x emissions; the formation of contrails; and the enhancement of cirrus cloud coverage. Although some of the impacts of aircraft on climate are well understood, others are not due to the complexity of the issues, the large scientific uncertainties that are associated with some processes, and the effects of trade-offs between different climate – and other environmental – effects. The IPCC (1999) report, *Aviation and the Global Atmosphere*, represented a landmark study of the impacts of air transport on climate. Since its publication, ongoing advances have been made in characterising the effects of aviation on the atmosphere. Further research is required to reduce the scientific and other uncertainties, to better inform decision-makers and to improve the understanding of the social and economic issues associated with the demand for air transport (IPCC 1999).

Yet despite the remaining scientific uncertainties, enough is known about the impacts of air transport on climate to indicate that, if current trends continue, the emissions from air transport could negate most or all of the emissions reductions achieved by other sectors of national economies. Thus the sustainable development challenge facing the air transport industry is brought to the fore: the imperative to address the impact of aircraft and airport operations on climate – and on other environmental systems – without blighting the economies and societies that depend upon the growth that the industry supports. That task presents a formidable challenge to policymakers, air transport industry representatives and researchers – especially because of the projected rapid growth of demand for air transport, the high abatement costs of the sector and the limited potential for radical technological or operational solutions to be found in the short to medium term. Consequently, the need to devise and implement an effective policy response is now urgent. Currently, the most promising approach appears to be the full inclusion of aviation CO_2 emissions within established emissions trading schemes (such as the EU ETS), although there are some contentious issues associated with that approach – especially the need to address the non-CO_2 impacts of aircraft on climate. The complexity and differing political acceptability of the various policy options means that no single instrument appears to be ideal, and the use of a combination of regulatory, market-based and voluntary approaches will probably be required in policy frameworks designed to reduce the impact of air transport on climate.

National commitments to achieve reductions in greenhouse gas emissions (which require dramatic emission reductions to be achieved by 2050) are likely to have profound implications for entire economies and societies (IPCC 2007; Stern 2007). Meeting such commitments is likely to require fundamental changes in the distribution and use of energy; in the development and availability of fuels; in infrastructure, business models, technologies and operating practices; and in the ways in which services are delivered. The air transport industry faces an immense challenge in adapting its activities to this changing context and the issue of climate change is likely to dominate debates about aviation environmental impacts and about the growth of the industry for many decades. Whilst, in the short to medium term, policy approaches are likely to focus on the inclusion of international aviation within emissions trading schemes and on facilitating voluntary agreements within the industry (involving carbon offsetting and commitments to achieve carbon neutrality), a more substantial response will be required in the long term. More contentious measures to limit the impact of air transport on climate include the imposition of more stringent emissions limits; the removal of existing privileges and subsidies of the industry; the wider use of emissions charges, fuel taxes and other levies; and ultimately, the use of severe demand restraint measures. Such measures would be extremely unpopular; hence, in the long term, the development of the air transport industry depends above all on finding radical technological solutions to reduce the impact of air transport on climate.

4 Air Transport and Air Quality

Introduction

Some of the earliest concerns about aviation environmental impacts emerged in relation to the effect of aircraft on air quality (ICAO 2007; Lee and Raper 2003, 77; Lee 2004, 2; see Chapter 1). Air quality is degraded by air pollution, the condition in which substances emitted by anthropogenic activities occur at elevated concentrations and produce a measurable effect on humans, animals, vegetation or materials (Seinfeld and Pandis 2006, 21). As explained in Chapter 2, aviation gives rise to a range of emitted species, most of which cause air pollution (Price and Probert 1995). In the past, the effect of aircraft and airports on air quality has been regarded a localised issue that is significant only in the immediate vicinity of airports and beneath their arrival and departure flight paths. However, recently, scientists have acknowledged that a continuum of air quality exists across the surface of the Earth; whilst human activities – including air transport – create 'hotspots' in which air pollution is concentrated, some pollutants may also have effects at considerable distances from the source of the emissions (Seinfeld and Pandis 2006; UK Air Quality Archive 2008). This is the case with air transport, which produces highly localised air pollution in the vicinity of airports as well as more widespread effects elsewhere (Hume and Watson 2003). Whilst many aviation emissions degrade air quality, the substances of greatest concern for air quality management are nitrogen oxides (NO_x) and particles. Emission levels of NO_x and particles may be critical during the landing and take-off (LTO) cycle, because those substances may cause air quality standards to be exceeded and may in turn constrain airport growth. Surface activities at airports – including the transport of employees, passengers, baggage and freight to, from and within airport sites – may also create substantial air pollution; at some airports, the localised air pollution due to aviation-related surface transport may be more significant than that generated by aircraft. Overall, aviation-related pollution is expected to increase alongside the projected rapid growth of air transport because the growth of demand for air travel is far outpacing the rate of technological and operational improvements in airframe and engine performance (see Chapter 2). Yet air quality standards reflect public concerns about the effects of pollution and those standards tend to become more stringent with time. Therefore, the impact of air transport on air quality is likely to remain an important issue on the decadal timescale, at least at the local level.

This chapter provides an account of the effects of air transport on air quality. It begins with an overview of the concepts of air quality and air pollution, providing some basic definitions and introducing key ideas such as air quality standards (or objectives) and air quality management. Some of the scientific techniques used to understand air quality and air pollution are outlined, including the use of emissions inventories, dispersion modelling and source apportionment (see Chapter 2). Subsequently, this chapter focuses specifically on the impact of air transport on air quality. A description of the main aviation-related emissions has already been provided in Chapter 2; below, the implications of those emissions for air quality are explored in more detail. In particular, the significance of aviation-related NO_x and particle emissions is explained. Next, this chapter covers a range of options for reducing the impact of aviation on air quality. As in the previous discussion of climate change (Chapter 3), the possibilities for mitigating the impacts of aviation on air quality include various technological, operational and policy options, all of which involve reducing emissions, primarily through the more efficient combustion of fuel. Technological options – such as improvements in airframe and engine design and in fuels – have formed the mainstay of efforts to reduce the impact of aircraft on air quality and considerable progress has been made in reducing specific emissions from aircraft engines. Operational options – including innovative methods of configuring, loading and manoeuvring aircraft as well as revised air traffic management (ATM) systems and procedures – have not yet been widely used to reduce air pollution, although they are frequently adopted as a means of conserving fuel, so they have the side-effect of reducing air pollution. However, again, as with the issue of climate change, those technological and operational improvements are being outstripped by the pace of aviation growth, with the result that effective policy is required to manage the impacts of increasing absolute emissions on air quality. A wide range of policy instruments could potentially be employed to manage those impacts; however, very few – other than the use of the International Civil Aviation Organization (ICAO) engine certification standards – are actually used. Consequently, at most airports, air pollution from aviation is not directly regulated. Nevertheless, air quality considerations play an increasingly important role in determining the outcome of planning applications for airport infrastructure developments; thus they may constrain aviation growth, both at airport level and more generally. Reducing the impact of aircraft on air quality remains a vital goal in its own right as well as an important part of the larger task of promoting sustainable development.

Air Quality

The atmosphere contains a mixture of gases together with a variety of liquid and solid particles. The proportions of those atmospheric constituents vary over time and space, and they are affected by both natural and anthropogenic processes (Kemp 2004). Clean air is a vital component of well-being, quality of life and, ultimately, life expectancy; the availability of clean air to breathe is increasingly regarded as a fundamental human right (DEFRA 2007b; DETR 1999; 2000). Air quality is influenced by the extent and severity of air pollution. In turn, air pollution is dependent on the types and concentrations of pollutants in the atmosphere, as the following definition suggests:

> A condition of 'air pollution' may be defined as a situation in which substances that result from anthropogenic activities are present at concentrations sufficiently high above their normal ambient levels to produce a measurable effect on humans, animals, vegetation, or materials. (Seinfeld and Pandis 2006, 21)

As Seinfeld and Pandis (2006) have acknowledged, this definition could include any substance, however noxious or benign, although the effects of air pollution are generally regarded as negative. Traditionally, air pollution has also been regarded as a feature of large urban areas and industrialised regions; however, scientists now recognise that the pollution associated with densely populated urban centres simply forms 'hotspots' in a continuum of trace species concentrations over the entire surface of the Earth (Seinfeld and Pandis 2006).

Air pollution is not a new environmental issue; it has occurred for millennia (Kemp 2004; Simmons 1998). For instance, airborne deposits of lead from Roman smelting activities have been detected in samples of Arctic ice (DETR 1999). Nor are public concerns about air quality and air pollution new. Historically, in developed and rapidly industrialising countries, air quality deteriorated dramatically with industrialisation, especially in major urban centres. The main air quality issue associated with industrialisation was the copious smoke and sulphur dioxide (SO_2) emissions that resulted from the combustion of sulphur-containing fossil fuels, especially coal. From the mid-nineteenth century, the operation of coal-fired processes in industrialised countries caused the atmosphere near major industrial cities to be frequently polluted by coal smoke, which gave rise in winter to the mixture of smoke and fog known as smog (DEFRA 2007b; DETR 1999). Complaints about the effects of air pollution accompanied industrialisation until, eventually, the introduction of early air pollution legislation in the 1950s and 1960s led to dramatic reductions in emissions from industrial and domestic sources. In the UK, the notorious Great Smog of 1952 that formed over London prompted the introduction of air quality legislation to protect human health from the impacts of air pollution. The UK Government introduced the *Clean Air Act 1956*, which was intended to control domestic sources of smoke through the introduction of zones in which only smokeless fuels could be burned. Whilst that legislation focused on reducing smoke pollution, the introduction of cleaner coals, together with the increased use of electricity and gas supplies, had the additional benefit of reducing sulphur dioxide (SO_2) emissions. Subsequent UK legislation, the *Clean Air Act 1968*, introduced a requirement for tall chimneys to be used in industrial processes in order to promote the wider dispersal of SO_2 emissions. Similar air quality legislation was enacted elsewhere, including, in the USA, the 1963 *Clean Air Act*, the first modern environmental law enacted by the US Congress (Seinfeld and Pandis 2006). The effects of air quality legislation, together with the increased generation of electricity in large, rurally-sited power stations, brought dramatic reductions in urban air pollution. Since the introduction of the early air quality legislation, most industrial and domestic pollutant emissions, together with their impacts on air quality, have tended to be either constant or decreasing over time, and industrial pollution has now ceased to be the dominant source of air pollution in urban areas within developed countries. Instead, the emissions from transportation sources – especially from road vehicles – now represent the major threat to air quality. Overall, recent trends in globalisation, mobility and demand for transport and tourism mean that pollutant emissions from transportation vehicles are now increasing at the global scale (UK Air Quality Archive 2008; see Chapter 2).

Air quality is an important concern because air pollutants have a range of negative environmental and social effects. A vast range of pollutants is produced by human activities, but the majority of air pollution due to transportation vehicles occurs in the form of emissions of NO_x, particles, hydrocarbons (HCs) and carbon monoxide (CO), which may significantly degrade urban air quality (UK Air Quality Archive 2008). In addition to those primary pollutants, ozone (O_3) is a secondary pollutant produced when primary emissions (for example, NO_x emissions) undergo photochemical reactions in the atmosphere; O_3 production can influence rural air quality at relatively large distances from the original source of the primary pollutant. Both primary and secondary air pollutants can affect human health, wildlife and vegetation in multiple, complex ways. NO_x emissions comprise nitric oxide (NO) and nitrogen dioxide (NO_2), but NO is rapidly converted to NO_2 in the atmosphere. At relatively high concentrations, NO_2 causes inflammation of the respiratory tract, affects lung function and exacerbates the response to allergens in sensitised individuals (CSP 2006; DEFRA 2007b). In addition, both NO and NO_2 are absorbed by vegetation, and exposure to high concentrations of those gases damages plant tissue and reduces growth rates (DEFRA 2007b; DETR 2000). A further consequence of NO_x emissions is the formation of acid precipitation, which leads to various – and sometimes severe – forms of ecological damage (Kemp 2004). Particles (also termed 'particulate matter' or 'particulates') are typically defined by their aerodynamic diameter (where PM_{10} refers to particulate matter of 10 micrometres or less in diameter and $PM_{2.5}$ refers to smaller particles of 2.5 micrometres or less in diameter). Emitted particles have a range of health implications: they affect the cardiovascular and respiratory systems, they exacerbate asthma and they cause a direct increase in mortality. The health effects of particles may be especially severe in people with pre-existing heart and lung diseases (DEFRA 2007b; DETR 2000; Stedman et al. 2007). The term 'hydrocarbons' (HCs) covers a diverse group of organic chemicals including benzene, polyaromatic hydrocarbons, kerosene, diesel, and de-icing compounds (such as ethylene glycol), some of which are acknowledged to be genotoxic carcinogens (DETR 2000; Hume and Watson 2003). Emissions of CO can be toxic to humans due to the formation of carboxyhaemoglobin, which reduces the oxygenation of blood and tissues and which poses a particular risk to individuals with pre-existing cardiovascular or respiratory diseases (DETR 2000). The secondary pollutant O_3 causes irritation to the eyes and damage to the respiratory tract, and also triggers inflammatory responses (CSP 2006; DEFRA 2007b; DETR 2000). In addition to these key substances of concern, a vast range of other air pollutants has been identified (Seinfeld and Pandis 2006). Overall, in Europe, poor air quality is responsible for severe impacts on human health, including around 370,000 early deaths in 2000 (EC 2005; Health Council of the Netherlands 1999; Netcen 2006). It is worth emphasising that carbon dioxide (CO_2) – whilst being an important greenhouse gas that contributes to climate change – is not regarded as a pollutant that degrades air quality or that directly affects human health.

The hazardous nature of many air pollutants has prompted the development of various air quality management frameworks. Such frameworks typically involve the establishment of air quality standards (or objectives) for key pollutants, together with the use of standardised air quality assessments and monitoring programmes, to ensure that the concentrations of those pollutants remain below specified limits – or at least that the occasions on which those limits are exceeded are kept to a minimum. The first such management framework in Europe, the UK National Air Quality Strategy (AQS), was published in 1997; it outlined specific commitments to achieve air quality objectives throughout the UK by 2005 (the UK AQS has subsequently been updated; AEA Technology Environment 2004; DEFRA 2007b; DETR 1999; 2000). In addition to setting national air quality objectives, the UK AQS also focused on areas of poor and declining air quality in order to reduce any significant risks to human health and to achieve broader sustainable development objectives relating to air quality. Thus the UK AQS acknowledged that national-level policies alone might not be sufficient to improve air quality in some areas. Areas in which air quality standards were exceeded (or were likely to be so) were designated as Air Quality Management Areas (AQMAs) in which local authorities were responsible for assessing and improving air quality. Subsequently, at the European level, the European Union (EU) has developed extensive legislation which sets health-based standards and objectives for a number of key air pollutants (EC 2008). Those standards and objectives apply over varying time periods because the observed health impacts associated with various pollutants occur over different exposure times. Of the air pollutants regulated by the EU, NO_2 and particles are considered to be the critical pollutants in terms of their effects on air quality in urban areas and around airports (DfT 2006; EC 2008; EEA 2008). The EU limit values for NO_2 are due to become mandatory from 2010, although the European Commission recently suggested that, under certain circumstances, the deadline for compliance with the annual limit values could be extended to 2015. In the US, the Environmental Protection Agency (EPA) developed National Ambient Air-Quality Standards (NAAQS) for six 'criteria pollutants': SO_2, particles, NO_2, CO, O_3 and lead. Under the US *Clean Air Act*, each state is required to adopt a State Implementation Plan (SIP) which allows for the implementation, maintenance and enforcement of the NAAQS (Kemp 2004; Seinfeld and Pandis 2006). Consequently, in some places, technological progress and the introduction of relatively strict environmental legislation have led to substantial improvements in air quality. Yet achieving further improvements in air quality presents a major challenge. Air quality management frameworks depend critically upon reducing emissions, a task that in turn requires the use of various policy instruments: regulations, emission limits, emission charges, taxes, subsidies, tradable permits and voluntary approaches. However, whilst those various policy instruments may be employed to manage air quality at the national level, very few have yet been used to manage the impacts of air transport on air quality, as subsequent sections will explain.

Scientific approaches to understanding air quality and air pollution have focused on two main types of technique: monitoring and modelling. In particular, those techniques have come to play a central role in informing decisions about planning approval for major infrastructure developments, including airport developments; hence it is crucial that reliable and accurate estimates and projections of emissions are available to decision-makers. Monitoring of ambient air quality involves the chemical analysis of air samples made under standardised conditions. In its simplest form, passive air quality monitoring may be undertaken using diffusion tubes containing two stainless steel gauzes and an absorbent trap; the tubes are left open (inverted) to the atmosphere but are sealed before being transported to the laboratory for analysis. The use of diffusion tubes has two main limitations: (a) they provide indicative, rather than highly accurate, data; and (b) they require a relatively long exposure period – typically of several weeks – and hence the results cannot be compared directly with air quality standards and objectives involving shorter averaging periods, such as hourly mean values. More accurate air quality monitoring techniques involve the use of automatic continuous monitoring sites at which routine *in situ* measurements of nitric oxide, nitrogen dioxide, particles and other species are made. At many locations, data are obtained using chemiluminescence devices with molybdenum converters installed at permanent air quality recording stations. Such monitoring stations must be rigorously and regularly calibrated and validated, and they typically provide hourly averaged concentrations for specific pollutants (Peace et al. 2006). More detailed, sporadic assessments of pollutant levels are also made using gas chromatography, spectroscopy and Light Detection and Ranging (LIDAR) techniques, although such measurements tend to be associated with individual research initiatives rather than with on-going air quality monitoring campaigns (Schürmann et al. 2007).

Monitoring approaches are complemented by modelling techniques, which can provide broader information about air quality for locations that are not served by monitoring stations and which can also be used to predict longer-term trends in pollution levels, although they also necessarily involve the use of simplifying assumptions. The primary tool used in modelling emissions and their geographical distribution is the emissions inventory: a database of all of the significant sources of a given pollutant in a particular area. Emissions inventories should ideally be compiled using a standardised methodology in order to ensure that they provide consistent, comparable data, although this is not currently the case. For a given airport, an emissions inventory should include all significant airside and groundside pollution sources as well as any other emission sources in the vicinity, even those that are not related to the operation of the airport. In air quality models, the emissions inventory is typically combined with a Geographic Information System (GIS) in order to allow the visualisation of the location and intensity of emissions from various sources. Since airborne pollutants are transported in the atmosphere as a result of dispersion processes, air quality modelling also involves the use of advanced dispersion modelling techniques to characterise the movement of emitted species during their expected atmospheric lifetimes (Peace et al. 2006; Seinfeld and Pandis 2006). It is important to emphasise that, for large and complex emission sources, such as airports, modelling studies should incorporate source apportionment techniques: they should investigate the source contributions to air pollution within an airport site rather than simply considering the airport as a single, homogenous source (Peace et al. 2006). Recent research in air quality modelling has focused on comparing different dispersion models and dispersion modelling techniques, on the uncertainties in characterising emissions and on the effectiveness of potential emission reduction scenarios.

The Impact of Air Transport on Air Quality

Air transport affects air quality, like climate, in several ways. Its impact can be divided into the effects of aircraft emissions, which represent the major contributor to aviation-derived air pollution, and the lesser but nonetheless significant air pollution that occurs due to the operation of airports and their ancillary services. As with the issue of climate change (discussed in Chapter 3), most scientific studies of this topic have focused on aircraft emissions rather than on emissions from airports, since the air pollution due to airport operations occurs in a similar manner to that generated by other large commercial sites. Again, the impact of air transport on air quality can be divided into direct and indirect effects (Table 4.1). Direct effects include emissions of air pollutants directly into the atmosphere, especially of NO_x and particles. In contrast, indirect effects occur when aircraft emissions initiate other physical or chemical processes that in turn cause air pollution, as when tropospheric ozone (O_3) is formed as a side-effect of NO_x emissions from aircraft and other airport-related sources. Aviation air pollutants may also be categorised as either gaseous pollutants (such as NO_x and CO) or aerosols (such as soot particles). A general overview of emissions has already been provided in Chapter 2, but the effects of a range of aircraft emissions on air quality are explained in more detail below and additional details of the implications of those effects are provided. In the account that follows, emphasis is placed on the critical emissions of NO_x and particles. However, other air pollutants are also emitted in relatively small quantities by aviation sources; those pollutants include sulphur oxides (SO_x), CO and volatile organic compounds (VOCs), and those species are also considered below (Schürmann et al. 2007). It is worth acknowledging that, whilst the main pollutants derived from aircraft have been well-documented, the composition of the aircraft exhaust plume is not yet fully known, and research programmes (such as the development of the ALFA aircraft plume analysis facility) are underway to determine the characteristics of aircraft emissions in greater detail.

Table 4.1   The main effects of aircraft on air quality

Effect	Significance
Emission of nitrogen oxides (NO_x)	Nitrogen oxides include nitrogen dioxide (NO_2), which has acute and chronic effects on human health, particularly in individuals with asthma, as NO_2 may cause inflammation of the airways and may affect lung function; in addition, NO_x enhances tropospheric ozone (O_3), a respiratory irritant; NO_x also has adverse ecological effects
Emission of particles (particulate matter, PM)	Particles are associated with a variety of health issues, including effects on the cardiovascular and respiratory systems, asthma and direct mortality
Emission of sulphur oxides (SO_x)	Sulphur oxides include sulphur dioxide (SO_2), which causes constriction of the respiratory airways, especially in individuals with asthma and chronic lung diseases; however, aircraft SO_x emissions are currently small
Emission of carbon monoxide (CO)	Carbon monoxide reduces the oxygen-carrying capacity of the blood, presenting a particular risk to individuals with pre-existing respiratory or cardiovascular diseases however, aircraft CO emissions are currently small
Emission of volatile organic compounds (VOCs)	Emissions of VOCs from aircraft include known genotoxic carcinogens (such as benzene and 1,3-butadiene) and are generally poorly understood; in addition, VOCs enhance tropospheric ozone (O_3), a respiratory irritant

Nitrogen oxides (NO_x), which consist of both nitric oxide (NO) and nitrogen dioxide (NO_2), are created in the combustor of the aircraft engine as a result of the oxidation of atmospheric nitrogen at high temperatures, with additional small quantities of NO_x being derived from the nitrogen content of the fuel (Rogers et al. 2002). As outlined in Chapter 2, most of the NO_x emitted by aircraft is in the form of NO, but this is rapidly converted to NO_2 in the atmosphere. However, a small proportion of NO_2 ('primary NO_2') is emitted directly from aircraft engines (Lee and Raper 2003, 84). Thus the effects of the NO_x emitted by aircraft are both direct (due to primary NO_2) and indirect (as a result of the conversion of NO to NO_2 in the atmosphere; DfT 2006). Whilst aircraft and other airport sources emit NO_x, the key air pollutant of concern is actually NO_2 due to its acute and chronic effects on human health, particularly in individuals with asthma. At relatively high concentrations, NO_2 causes inflammation of the airways, and long-term exposure to NO_2 may affect lung function and may exacerbate the response to allergens in sensitised individuals. Both hourly mean and annual mean objectives have been set for NO_2 (DEFRA 2007b; DETR 2000). Given that all combustion processes in air create NO_x, the production of NO_x during the operation of aircraft engines is unavoidable, although NO_x emissions can be reduced by careful combustor design (Lee 2009; Rogers et al. 2002). Nevertheless, due to the fact that NO_x production tends to increase as an unintended consequence of the drive to achieve greater fuel efficiency, absolute NO_x emissions from aircraft are forecast to increase by a factor of approximately 1.6 between 2002 and 2025 (Eyers et al. 2004). Consequently, doubts about whether airport expansion could result in local air quality standards for NO_2 being exceeded at major airports have generated considerable concern – as, for instance, in relation to proposals to develop a third runway at London Heathrow Airport (DfT 2006). Aviation-related NO_x emissions are most concentrated in the vicinity of the runways, taxiways and aprons of airports, as well as close to major roads, and recent work has demonstrated that exhaust pollutants may be transported to the ground far more effectively by aircraft wakes than by ambient atmospheric dispersion alone (Graham and Raper 2006a; 2006b; Peace et al. 2006; Schürmann et al. 2007; Underwood et al. 2001).

NO_x emissions have other, secondary effects on air quality besides the primary effects of NO_2: specifically, the formation of O_3 and particles (DETR 2000). Both NO_x and VOCs undergo complex chemical reactions resulting in the catalytic production of tropospheric O_3 (Brasseur et al. 1998; DETR 2000; IPCC 1999; Lee 2004; 2009; Lee and Raper 2003; Rogers et al. 2002). The effects of tropospheric O_3 on climate have been discussed in Chapter 3; however, in addition to those effects, O_3 plays an important role in converting NO (the predominant component of NO_x) to NO_2 (DfT 2006). O_3 is also an important air quality pollutant in its own right because it can cause irritation to the eyes and nose, damage to the lining of the respiratory tract, and inflammatory reactions (DETR 2000). The photochemical reactions that generate O_3 are not instantaneous but occur over timescales ranging from several hours to several days. Consequently, a running eight-hour mean objective is typically set for tropospheric O_3 (DEFRA 2007b; DETR 2000; see also Lee 2004). Besides their role in the production of tropospheric O_3, NO_x emissions are also implicated in the formation of particles (see below).

Particles – which are alternatively known as 'particulate matter' (PM) or 'particulates' – are emitted by aircraft, airport stationary plant, ground support vehicles and passenger surface transport vehicles. The particles emitted by aircraft are diverse: they are emitted from (carbon) brake and tyre wear as well as from main engines and auxiliary power units (APUs) (Hutton et al. 1999; Underwood et al. 2001). In addition, substantial particle emissions may occur due to the operation of diesel-powered surface vehicles. Soot and sulphate particles form a significant fraction of aviation particle emissions; those substances form aerosols – suspensions of fine solid or liquid particles in a gas – in the atmosphere (Seinfeld and Pandis 2006). The aerosols may be emitted directly as particles ('primary aerosol') or may form in the atmosphere as a result of conversion processes ('secondary aerosol'). Typically, atmospheric aerosols are regarded as those containing particles ranging from several nanometres to tens of micrometres in diameter (Seinfeld and Pandis 2006). Conventionally, particulate matter has been categorised as particles with an aerodynamic diameter of 10 micrometres or less (PM_{10}) or of 2.5 micrometres or less ($PM_{2.5}$). In fact, most of the particles emitted by aircraft are $PM_{2.5}$, which are of greater concern than larger particles in terms of their effects on human health. Particles are associated with a variety of health issues, including effects on the cardiovascular and respiratory systems, asthma and direct mortality. The health effects of particles tend to increase with the concentration of the pollutant and are likely to be especially severe amongst people with pre-existing heart and lung conditions (DETR 2000). Both hourly mean and annual mean objectives have been defined for PM_{10} (DEFRA 2007b; DETR 2000). In addition, in Europe, new $PM_{2.5}$ objectives have been introduced in order to mitigate human exposure to very fine particles. Those objectives are based on the use of the average exposure indicator (AEI), which is defined as a three-year running annual mean $PM_{2.5}$ concentration averaged over selected monitoring stations in large urban areas and agglomerations (EC 2008).

Scientific assessments of particle emissions from aviation sources are characterised by greater uncertainty than are those of NO_x emissions; however, the contribution of airport sources to ambient concentrations of particles in areas adjacent to airports tends to be smaller than that of airport-derived NO_x (DfT 2006). The highest concentrations of aviation-derived particles occur close to runways, taxiways, aprons and major roads; furthermore, as with NO_x, particles can be very effectively transported to the surface by aircraft wakes (Graham and Raper 2006a; 2006b; Peace et al. 2006). The development of airport infrastructure could potentially be constrained if the expansion of a given airport could result in local air quality standards for PM_{10} or $PM_{2.5}$ being exceeded (DfT 2006). Given that measures can be taken to ensure that particle emissions are minimised from many airport sources – for example, through the conversion of ground support vehicles to burn liquefied petroleum gas (LPG) fuel instead of diesel – the air quality objectives for particles may represent less of a constraint at most airports than do the objectives for NO_x. Nevertheless, particle emissions are forecast to increase over the period 2002–2025 due to the difficulty of sustaining significant technology improvements in this area, meaning that aviation-derived particles are likely to remain an air pollutant of critical concern over the decadal timescale (Eyers et al. 2004). Furthermore, in some countries, particle emissions from road traffic are successfully being reduced with the result that aviation sources of particles are responsible for an increasing share of the total particle emissions in those countries.

Other local air pollutants emitted by aircraft include SO_x, CO and VOCs; those substances are also emitted by other airport sources besides aircraft (Price and Probert 1995; Schürmann et al. 2007). Emissions of SO_x occur from aircraft because sulphur-containing compounds are added to aviation fuel to improve its lubricity; SO_x is also emitted from surface transportation vehicles and other airport sources, although those vehicles increasingly use low-sulphur diesel or alternative fuel or energy sources. SO_x emissions include the key air pollutant, sulphur dioxide (SO_2), which may cause constriction of the respiratory airways, especially in individuals with asthma and chronic lung diseases. Since the health effects of SO_2 occur almost instantaneously upon exposure to the gas, the SO_2 standards typically include hourly mean, 24-hour mean and 15-minute mean objectives (DEFRA 2007b; DETR 2000). However, aircraft SO_x emissions are generally very small and they rarely present significant air quality issues at airports (DfT 2006). Emissions of CO occur as a result of the incomplete combustion of fossil fuels; CO presents a significant risk to human health due to its role in the formation of carboxyhaemoglobin, which reduces the oxygen-carrying capacity of the blood and which presents a particular risk to individuals with pre-existing respiratory or cardiovascular diseases. Typically, a running eight-hour mean is used as the CO objective (DEFRA 2007b; DETR 2000). However, the high conversion efficiency of modern turbofan engines means that very little CO is generated by commercial aircraft operations, although a greater proportion of CO may be emitted by other airport sources such as airside service vehicles (Yu et al. 2004). Nevertheless, the aviation-related sources of this pollutant are generally small (DfT 2006). Therefore, whilst SO_2 and CO are significant pollutants in general terms, they are not regarded as major outputs of aircraft operations and they do not represent issues that currently constrain aviation growth; nor can the health impacts of aviation-derived emissions of these pollutants readily be distinguished from those occurring under ambient conditions (Hume and Watson 2003; Yu et al. 2004). In contrast to SO_x and CO, however, emissions of VOCs from aviation sources may represent a greater concern since they include known genotoxic carcinogens such as benzene and 1,3-butadiene – and because emissions of HCs from aircraft are generally poorly understood (Anderson et al. 2006; DETR 2000, 39; DfT 2006). A recent study by Schürmann et al. (2007) indicates that, among the VOCs, reactive C2-C3 alkenes may be found in significant quantities in aircraft engine exhaust compared to ambient levels; those authors also found that another VOC, isoprene, is found in aircraft engine exhaust (although not in refuelling emissions). Another reason why emissions of VOCs (as well as of NO_x) give cause for concern in relation to air quality is their secondary effect in the formation of another pollutant, tropospheric O_3 (see above).

The discussion above has focused primarily on the air quality impacts of emissions from aircraft, which represent the majority of aviation-related effects on air quality. However, impacts on air quality also occur due to the operation of airports and their ancillary services (DfT 2006; Hume and Watson 2003). The construction, operation and maintenance of airports results in the emission of gaseous pollutants and aerosols due to energy conversion and supply, the production and transport of materials for consumption within airports, the combustion of fossil fuels and the disturbance or destruction of vegetation. Gaseous pollutants and aerosols are emitted by surface transport vehicles used by passengers and airport staff to travel to and from airports. Further emissions occur due to the provision of services at airports, including baggage handling, catering, cleaning, refuelling and waste management. Airside surface vehicles that are powered by diesel or petrol fuel may also make significant contributions to gaseous and aerosol emissions at airports, although airport surface vehicle fleets are increasingly being renewed with vehicles that use low-sulphur diesel, alternative fuels or other energy sources. Whilst airport emissions and their impacts on air quality are small in comparison with aircraft emissions, they may nonetheless be cumulatively significant, especially in areas where compliance with air quality standards is already marginal – as is often the case in the vicinity of major airports located near to large urban centres. As air quality objectives in the vicinity of airports become increasingly stringent, airport operators are obliged to work more closely with local authorities in order to assess and manage impacts on air quality. In turn, other parts of the air transport industry – especially airlines – are increasingly expected to engage with the process of improving air quality by reducing the emissions arising from their operations. In comparison with aircraft emissions, the air quality impacts of airports and their associated surface activities may be relatively easy to manage, although many complexities and uncertainties remain in the characterisation of those emissions (Peace et al. 2006; Schürmann et al. 2007; Unal et al. 2005; Winther et al. 2006).

In the previous discussion, the main aviation-derived substances that affect air quality have been described. However, knowledge of the main pollutants emitted by aircraft and airport sources is not sufficient to fully characterise aviation-derived air pollution or its impacts on air quality. This is because air pollution is generally a complex phenomenon that is unlikely to be characterised adequately as a given concentration of a pollutant dispersed uniformly in a given volume of air. The distribution of pollutants in the atmosphere – and particularly in the turbulent atmospheric boundary layer adjacent to the Earth's surface – is rarely homogenous and the emission of pollutants varies spatially and temporally. Pollutant sources vary: they include point and area sources, and (as in the case of aircraft) they may be either moving or stationary (Peace et al. 2006; Stedman et al. 2007). The dynamics of pollutant dispersion – and the varying atmospheric lifetimes of different pollutants – also complicate the picture of air pollution. Pollutants are entities that may interact with each other, and with other atmospheric species, both chemically and physically. Moreover, the medium into which pollutants are discharged (the atmosphere) is itself constantly changing and difficult to characterise accurately. Given such complications, advanced monitoring and modelling techniques are required in order to provide a detailed picture of the air quality at airports and, ideally, to create projections of the consequences of specific airport development scenarios for air quality. In the previous section, some of the general principles of the scientific monitoring and modelling of air pollution were briefly described. Recently, some important studies have been undertaken specifically to investigate the impacts of aviation emissions on air quality, and considerable progress has been made, although some important uncertainties remain. In particular, concerted efforts have been made by scientists to consider the effects on air quality of projected airport emissions for various years in the future, and for particular airports (DfT 2006; Kesgin 2006). In addition to focusing on air quality impacts associated with individual airports, advanced air pollution projections also take into account the total air pollution levels from all sources and the source contribution (using source apportionment techniques) of the main airport sources to those totals. Various issues emerging from those studies are discussed below.

In recent years, it has become clear that the accurate calculation of aviation emissions – and their effects on air quality – depends in the first instance upon obtaining a reliable understanding of all of the significant sources of air pollutant in the vicinity of a given airport. In that task, the primary tool is the emissions inventory (Peace et al. 2006). Whilst a standardised methodology for compiling airport emissions inventories has yet to be developed, some consensus exists about the characteristics of an effective inventory. Given the complexity of the sources of pollution at major airports and the wide variety of data required, the task of compiling a comprehensive emissions inventory is a demanding and resource-intensive process. An emissions inventory should distinguish between different sources of air pollution at an airport: aircraft, surface vehicles (both airside and landside), fuel storage, refuelling operations, and other stationary sources (such as terminal boilers and heating plants). Ideally, once identified, those sources should be incorporated in a GIS, which allows the location and intensity of each emission source to be visualised. Thus in order to characterise the location and intensity of the emissions from a given source accurately, a geographical description of each source (for instance, point, line, area or volume sources) should be included in the model. A further advantage of linking the emissions inventory with a GIS is that spatially-resolved emissions data can function as the basic inputs for air quality dispersion modelling, which in turn allows the likely dispersion of air pollutants from any given airport source to be predicted and visualised. Furthermore, data from discrete airport sources may be aggregated to allow prediction of the dispersion pattern of a given pollutant from the entire airport site. Dispersion models may also use data describing meteorological conditions over a given time period in order to provide realistic predictions of the transport of pollutants in the atmosphere; if possible, dispersion models should be updated regularly using actual ambient meteorological data.

Ideally, an emissions inventory includes observed emissions data, measured using reliable, standardised techniques over a long time period (such as a year) in order to ensure the statistical significance of the output. However, it is not feasible to measure emissions from all airport sources over long periods. Hence the compilation of an emissions inventory involves the use of estimation methods, which are dependent upon the use of simplifying assumptions. Consequently, various uncertainties are associated with emissions inventories due to the use of estimation methods. First, the secondary data used may be incorrect; those data typically include information about activity rates at a given airport, such as total numbers of air traffic movements, surface transport usage, or frequencies of ground support activities. Second, the use of emission factors is required in order to estimate the quantity of a pollutant released for a given unit of activity (see Chapter 2). Those emission factors are generally provided by organisations such as the International Civil Aviation Organization (ICAO) or the FAA, and they are typically taken from databases of standard emission factors that may be based on a relatively small number of observations made under idealised conditions (such as at the point of engine certification). Those standard emission factors may differ substantially from the actual values occurring under operational conditions, for a wide variety of reasons (Schürmann et al. 2007; Unique 2004; 2005). Third, if attempts are made to incorporate actual operational data in an emissions inventory rather than standard emission factors, those attempts may be hindered by insufficient operational activity data. For instance, airport-specific 'time-in-mode' measurements may not be available, making it impossible to determine, say, the actual average time spent by each aircraft in taxiing to the runway, for each runway, and for a range of ambient conditions.

In compiling data for emissions inventories, aircraft exhaust emissions are generally calculated over one complete LTO cycle. This is because emissions due to aircraft operations above the upper limit of the LTO cycle (for example, during the cruise) do not significantly affect air quality at the Earth's surface. A standardised LTO cycle has been defined by ICAO as extending up to 3,000 feet (approximately 1,000 metres) above ground level (IPCC 1999). For that standardised LTO cycle, emissions for each particular airframe-engine combination may then be calculated using the emission factors for each aircraft's specific engines, at each power setting, for each mode of operation (such as take-off), using information about the time spent in each mode. In principle, the activity of the aircraft – in terms of time spent in each mode of operation – for the period covered by the inventory can then be used to estimate the total emissions produced by that aircraft. This procedure represents a standard approach to the calculation of aircraft emissions (EPA 1995; 1997). One limitation of this approach, however, is that it can only be used for those pollutants for which standard emission factors have been defined (such as NO_x, CO and HCs). Emissions of other pollutants, such as the evaporative emission of VOCs during aircraft refuelling, are not easily characterised and very limited information is available about their nature or extent – partly because emissions of VOCs during refuelling are thought to be small due to the low vapour pressure of the fuel and the use of quick-connect refuelling nozzles (FAA 1997). Nevertheless, the cumulative effects of VOCs released as a result of many refuelling operations might potentially be significant, yet those pollutants tend to be characterised poorly in emissions inventories (Schürmann et al. 2007).

To some extent, methods for obtaining emissions inventory data for other pollutants, including particles, are available. Emissions of particles may be estimated for the limited number of engines for which some data exist. However, additional research is required in order to develop accurate estimation methods of describing particle emissions at airports (EPA 1997). In the absence of more accurate data, a crude method of estimating particle emissions, based on the use of the ICAO smoke number, has been employed (Petzold 2001; Underwood et al. 1996). That method uses the empirically-observed relationship between particle emissions and smoke number; the latter values are available in the ICAO Aircraft Engine Emissions Databank (CAA 2006). Smoke number is a relative measure calculated on the basis of the reflectance of filter paper before and after the passage of a known volume of smoke-bearing samples. Whilst 'smoke number' and 'particle mass concentration' refer to different phenomena in relation to engine exhaust gas, a universal curve (the 'curve of Champagne') may be used to relate smoke number to particle concentration. Underwood et al. (1996) have estimated emission factors for particles based on this relationship; those authors assumed a single value smoke number for all engine types, with separate values selected for each thrust setting (and hence for each mode of the LTO cycle, since each mode is associated with a typical thrust setting). The approximation of particle emissions by this method generally provides conservative values, resulting in the over-estimation of actual particle emissions.

Nonetheless, in general, the annual emissions of a pollutant can be calculated for a given aircraft at a given airport, for each mode of operation, based on knowledge (or estimates) of the following variables: the number of aircraft movements; the fuel consumption during each mode; the time spent in each mode; and the emission factor for that pollutant. This calculation can be repeated for each aircraft, and for each pollutant, in order to compile an emissions inventory for that airport. Emission factors and average fuel consumption rates for aircraft engines are published for each of the main operating modes of the LTO cycle (approach, idle, take-off and climb-out) in the ICAO Aircraft Engine Emissions Databank (CAA 2006). The duration of each mode, for a particular aircraft, depends on various factors such as the aircraft mass, ambient meteorological conditions and pilot skill. To the extent possible, airport-specific data should be used in calculating the average duration of each operating mode (the 'time in mode') for each aircraft type. Since airport-specific data are not always available, ICAO has defined a standardised LTO cycle based on time in mode data averaged from a large number of international airports. As with time in mode data, the use of airport-specific data about the aircraft and engines actually operating there (including data about specific engine types and the numbers of engines per aircraft) will improve the accuracy of the emissions calculation. The aircraft-engine combinations actually operating at a given airport can be determined using the unique registration code of each aircraft, which may be cross-referenced with a database of airframe-engine types, such as the JP Airline-Fleets International Manual. It is important to use data about actual airframe and engine types in the emissions calculation, if possible, since there may be wide variations in performance even between apparently similar types (Schürmann et al. 2007). Whilst the emissions calculation method described above is generally used for the main aircraft operating modes, emissions from aircraft engine testing may also be calculated using a similar method, provided that information about the duration and type (idle or full power) of the test is available.

Other airport sources besides aircraft main engines should also be included in emissions inventories. Aircraft auxiliary power units (APUs) are kerosene-fuelled generators carried aboard aircraft, which provide power for the aircraft whilst it is on-stand. APUs may routinely be left running for considerable periods of time (for instance, during the pre-flight servicing of aircraft) and they may also operated during flight in preparation for landing. Therefore, APUs may contribute substantially to air pollution in the vicinity of airports, yet very few emissions data are available for APUs, especially for their particle emissions (although APU particle emission factors are generally higher than those of modern aircraft main engines; DfT 2006; FAA 1997). Airport service vehicles represent another emission source at airports; they are used in a wide range of service and supply functions at airports, both airside and landside. Emissions from airport service vehicles may be calculated using standard vehicle emission methodologies, such as those published in the DEFRA guidance on vehicle emission estimation. However, accurate data about both the numbers of vehicles operating and about the fuel consumed by those vehicles may not always be available. Ground support equipment also contributes to air pollution at airports; emission factors for ground support equipment, covering a range of fuel types, have been published by the FAA (1997), although those emission factors cover wide ranges of values and they contain considerable uncertainty. Further sources of air pollution at airports are fuel storage tanks and refuelling processes, due to evaporative emissions of HCs. Emissions from fuel storage tanks depend upon the tank type (fixed- or floating-roof) and dimensions, fuel type, ambient meteorological conditions and operating practices. In general, fixed-roof tanks produce more emissions because the vapour space in the tank becomes saturated with HC vapour which may then escape when the tank is refilled. Floating-roof tanks minimise this problem by eliminating the vapour space between the liquid level in the tank and the tank roof. The US EPA has published a methodology for calculating HC emissions from fuel storage tanks, covering most types of tank, in its *Compilation of Air Pollutant Emission Factors* (EPA 1995). Finally, other emissions at airports arise from the combustion of fuels in other airport-related processes, especially as a result of the stationary power plant used for electricity generation, heating and air conditioning. The emissions from such sources may be calculated using generic, widely-available emission factors.

Emissions inventories have been compiled for many airports, generally in association with infrastructure development proposals. For instance, an emissions inventory for London Heathrow Airport was produced as part of the planning application for its Terminal 5. Another was compiled for Manchester Airport as part of its second runway development proposal and it was presented at a Public Inquiry in 1995. Hence emissions inventories may be important tools in the strategic planning of airport development as well as in airport environmental management. Emissions inventories are useful instruments not only because they provide an understanding of the most significant sources of pollution at a given airport but also because they may allow comparisons to be made between airports. Thus current and future year analysis of air quality in the vicinity of 23 regional airports in the UK was undertaken by AEA Technology plc, a process that informed the UK Department for Transport's (DfT) *Regional Air Services Co-ordination Study* (DfT 2002). In that study, aircraft emissions were estimated using aircraft movement data provided by the UK Department of Environment, Transport and the Regions (DETR), and generic time in mode data and emission factors from the ICAO Engine Emissions

Databank were also used. The pollutant concentration estimates published in that study were based on conventional atmospheric dispersion modelling techniques.

Whilst recent work associated with the production of emissions inventories has provided some important insights into air pollution at airports, there is still a pressing need for further research into the impacts of air transport on air quality. Although recent research programmes have made progress in elucidating the composition and significance of aircraft emissions, better understanding of the complex physical and chemical processes occurring in the engine exhaust plume is required. Considerable scientific uncertainty remains about the precise composition of the aircraft exhaust in terms of its gaseous and particle components. Until recently, in Europe, there has been very limited capability to measure aircraft emissions in the dynamic environment of the exhaust plume due to the lack of suitable sampling equipment. However, recent initiatives, such as the development of the ALFA aircraft plume analysis facility, are in progress with the aim of understanding the composition and dynamics of aircraft plumes with unprecedented accuracy. Studies conducted using the ALFA facility are expected to focus on improving scientific understanding of the composition and dispersion of plumes and to contribute to a growing database of operational aircraft emissions. One outcome of this work is likely to be the more accurate characterisation of aircraft plumes during the LTO cycle, which could in turn inform better air quality assessments. In addition, whilst not being of direct relevance for air quality at the Earth's surface, an important side-benefit of plume analysis studies would be improved understanding of aircraft emissions during the cruise, which are responsible for high-altitude pollution. Plume analysis studies could also provide new insights into the effectiveness of operational measures designed to reduce the impacts of aircraft on air quality; those operational measures include reduced thrust take-off techniques and the use of modified fuels (see below). Yet, even with the potential insights offered by plume analysis studies, the task of characterising aircraft emissions more accurately remains extremely challenging. Jet efflux is a complex mixture of hot, high-speed gas and cooler, slower-moving gas; thus aircraft exhaust is heterogeneous and highly turbulent, and multiple scale and chemical reactions occur within the plume. Consequently, further research into aircraft plume characteristics requires the use of computational fluid dynamics (CFD) in order to construct accurate models of the flow immediately downstream of the exit of the engine, and of the subsequent mixing processes. Research is in progress to determine how the efflux from a jet engine is transformed into a mixed plume, as well as to provide further details of the composition of the plume itself.

Further research is also needed to clarify the way in which aircraft emissions disperse in the atmosphere. To date, sensitivity analysis of initial plume dispersion parameters has been undertaken; that analysis has identified the need for more detailed investigations of the initial dispersion characteristics of emissions, and of their parameterisation. Improved data are required to verify advanced dispersal models. In particular, better measurements of the dispersion and evolution of emissions from aircraft engines are required, both during ground manoeuvring and in flight. Recent research initiatives, such as the AETIAQ (Aviation Emissions and Their Impact on Air Quality) study, offer potential new insights into those dispersion and evolution processes. Improved data are also required in order to determine whether enhanced peak aerosol concentrations occur as aircraft-induced vortices dissipate near the ground in the vicinity of airports. During take-off and landing, the wings of an aircraft produce relatively high lift forces which in turn generate powerful trailing vortices; those vortices interact with the engine exhaust plumes and they modify the dispersal of the jet efflux. However, scientific understanding of this phenomenon is currently limited, and new insights into the interactions between vortices and exhaust plumes are required (Graham and Raper 2006a; 2006b). Another, recent approach to the study of the dispersal of aircraft emissions involves data collected using atmospheric boundary layer wind tunnels in which the conditions of aircraft engines in flight may be simulated – and exhaust plumes analysed – in the context of a variety of ambient wind conditions; such an approach could potentially inform new models of efflux dispersion. To date, this approach has been little used for simulating aircraft engine exhaust plumes, but research initiatives are in progress to assess the main factors influencing plume trajectory and concentration levels under various simulated wind conditions, and for a range of aircraft operations. Perhaps the most pressing area in which further research is required to clarify the impacts of aviation on air quality relates to particle emissions, which are still relatively poorly understood and for which real-time monitoring over a size range relevant for human health (0.1–10 micrometres aerodynamic diameter) is in its infancy (DfT 2006). There is an urgent need to understand the composition, numbers and sizes of particles emitted by aircraft, since such details are critical to understanding the effects of particles on human health, and thus in informing regulatory standards. A related research priority is to make improved techniques available to airports for the routine monitoring of air quality.

Reducing the Impact of Air Transport on Air Quality

As with the issue of climate change (Chapter 3), the possibilities for reducing the impact of air transport on air quality include a wide range of technological, operational and policy options; the distinctions between those various options are not always clear-cut. All of those options are based on the fundamental principle of reducing emissions, primarily through achieving more efficient fuel consumption, although the question of whether it is specific or absolute emissions that are reduced is a crucial one (see Chapter 2). In relation to aviation, approaches to air quality management focus on reducing emissions both from aircraft during the LTO cycle and from other sources at airports. However, in contrast to the issue of climate change (which, in relation to aviation, is dominated by the effect of aircraft emissions), air quality is affected by aircraft, airport and other (non-aviation) emissions in a more balanced way (DfT 2006). In relation to aircraft, technological options for reducing emissions focus on improvements in airframes, engines and fuels, whilst operational options potentially include the use of innovative methods of configuring, loading and manoeuvring aircraft, as well as revised ATM systems and procedures. In relation to other airport sources, a much wider range of options exists for reducing emissions, ranging from the provision of off-site car-parking facilities to the use of renewable energy to power airport terminal buildings. To date, the impact of air transport on air quality has not prompted the introduction or development of a wide range of policy options; indeed, few policy measures (besides the application of ICAO engine certification standards) are used to manage aviation impacts on air quality. Consequently, at many airports, aviation air pollution is not directly regulated but may be covered by general air quality standards that apply to the areas in which airports are located – areas that include a variety of other, non-aviation sources of pollution. Given that the impact of air transport on air quality – especially in the vicinity of large airports – is expected to increase due to the projected rapid growth of demand for air transport (and also because that growth is likely to be accompanied by increased demand for surface access, which could further increase aviation-related emissions), the need to implement effective policy options to reduce that impact is becoming more pressing, at least for major airports (Peace et al. 2006; Unal et al. 2005; Yu et al. 2004).

Reducing the impacts of air transport on air quality is not a straightforward task. As previously noted (Chapters 2 and 3), the air transport industry is international in its scope, is growing rapidly, is technologically mature and is characterised by long lead-in times and long in-service lifetimes. Whilst incremental improvements in fuel efficiency – with associated reductions in specific emissions of most pollutants – are expected to continue to occur due to refinements in airframe and engine designs, the potential to achieve more radical emission reductions through technological advances is much more limited, at least in the short to medium term (Stern 2007). Operational options have been little explored, to date, for the purpose of reducing the impact of air transport on air quality, although they are widely used to improve fuel efficiency and they probably reduce aircraft emissions as a side-effect. However, even under the most optimistic forecasts of rates of operational improvement – including advances in the efficiency of ATM systems and procedures – these are unlikely to offset the rise in absolute emissions associated with aviation growth. Therefore, effective policy is required to manage the intensifying impact of air transport on air quality at major airports. Particular concerns focus on the growing problem of NO_x and particle emissions in the vicinity of airports and, increasingly, on the little-known effects of aviation-derived VOCs (Schürmann et al. 2007). Consequently, the management of the impact of air transport on air quality involves seeking technological solutions to limit NO_x emissions whilst simultaneously sustaining improvements in overall fuel efficiency and reductions in CO_2 emissions (since a trade-off exists between reducing NO_x emissions on the one hand and minimising fuel burn and CO_2 emissions on the other; ACARE 2004). That trade-off underlines the point that, unless efforts to reduce NO_x emissions are coordinated with other environmental initiatives, they may undermine attempts to reduce greenhouse gas emissions (Chapter 3). If air quality objectives are not to constrain the growth of airports and of aviation more generally, policymakers face the task of finding technological, operational and policy solutions that will substantially reduce the impact of aircraft and airport emissions on air quality. Using the framework outlined by the IPCC (1999), the options for mitigation of that impact may be categorised in four main groups: (a) engine and airframe technology options; (b) fuel options; (c) operational options; and (d) regulatory, economic and voluntary options. Below, those technological, operational and policy options are examined in turn.

Engine and Airframe Technology Options

The main ways in which improved airframe and engine technologies may reduce the impact of aircraft on air quality are similar to those that could potentially be used to reduce climate impacts (Chapter 3). Those options are based on two main principles: (a) improving fuel efficiency, which has the effect of reducing specific emissions of most pollutants; and (b) improving the design of aircraft engines – and particularly of fuel combustors – in order to reduce specific emissions of NO_x and particles. As previously mentioned, technological advances have dramatically improved the environmental performance of airframes and engines and the specific emissions of most aircraft-derived pollutants have been reduced as a consequence of improved fuel efficiency (Thomas and Raper 2000; see Chapter 2). Incremental advances in engine performance are expected to continue to be made; the IPCC (1999) has stated that a projected 20 per cent improvement in fuel efficiency is expected to occur by 2015, and a 40 per cent improvement by 2050, relative to aircraft manufactured in 1999. Internationally, various major technology research programmes have been established with the aim of reducing NO_x emissions during the LTO cycle whilst simultaneously improving engine fuel efficiency by 2010 (ACARE 2004; IPCC 1999). Such improvements in aircraft engine technology could potentially make the difference between compliance and non-compliance with air quality objectives for NO_x in the vicinity of major airports under particular airport infrastructure development scenarios (DfT 2006).

Aircraft fuel efficiency can be increased by means of aerodynamic improvements, weight reductions and the use of more fuel-efficient engines and other systems (Bows et al. 2009; Drake 1974; RCEP 2002; see Chapter 3). Airframe improvements are based on the principle of reducing aerodynamic drag, which has the side-benefit of reducing most aircraft emissions as well as operating costs (Lee 2004). Such improvements include advances in the use of laminar flow technology, riblets, wing-tip devices, high aspect-ratio/low sweep configurations; more radical airframe designs, such as the blended wing-body and the wing-in-ground-effect designs, are also being investigated (Air Travel – Greener By Design 2005; 2006; 2007; 2008; Bows et al. 2009; Viswanath 2002). Approaches based on weight reduction include the use of lightweight materials (carbon fibre or reinforced plastic structures) in the construction of aircraft components (Finley 2008; Marsh 2007). Other approaches focus on the use of adaptive structures in the airframe, which could allow aircraft to be operated consistently at their design conditions, leading to more fuel-efficient flight; such adaptive structures could potentially also reduce the complexity and weight of aircraft by dispensing with additional control surfaces (ACARE 2004). Overall, however, airframe modifications are unlikely to yield large reductions in aviation emissions (Lee 2004). Developments in engine technology appear to be more promising: research programmes are investigating several key areas: combustor sub-component performance, complex multi-staged combustors, single-stage combustor technology, fuel injection, enhanced fuel-air mixing, combustor liner coolant flow and hot-gas residence time (Lee 2004). However, whilst NO_x emissions could potentially be reduced through the use of 'lean-burn' combustion techniques, such lean-burning, low emissions combustors are susceptible to combustion instabilities (Lee 2009).

The task of reducing NO_x emissions using technological means is not straightforward; it involves making trade-offs with the control of other pollutants, especially CO_2. Attempts to develop quieter and more fuel-efficient engines have prompted increases in the overall pressure ratio of engines. In general, under conditions of higher operating temperatures and pressures, more NO_x is generated, with the result that combustor technology must make commensurate progress in order to maintain or improve NO_x emission control (Lee 2004). This represents a major engineering challenge, particularly given that, ideally, aircraft engines would be optimised for minimal NO_x formation over a wide range of different engine operating conditions (Lee 2009). Thus the design of future engines and airframes involves a complex decision-making process and many considerations must be balanced, including CO_2 emissions, NO_x emissions at ground level, NO_x emissions at altitude, water vapour emissions, contrail formation, cirrus cloud production and aircraft noise (IPCC 1999). As with the issue of climate change, technological approaches to reducing the impact of air transport on air quality involve balancing a range of aviation environmental impacts and making rational trade-offs – based on scientifically-robust data and on agreed priorities – between the management of those impacts. Overall, advances in airframe and engine technologies are likely to assist in reducing the impact of aircraft on air quality; but, until radically new airframe or engine designs become available, the use of policy-based approaches will also be required (Peeters et al. 2009; RCEP 2002).

Of course, aircraft main engines are not the only engines operating at airports: APUs, surface vehicles and airport terminal power, heating and cooling plants also burn fuels and generate emissions of significance for air quality (DfT 2006; Peace et al. 2006). Technological improvements in the efficiency of those engines can also potentially improve air quality in the vicinity of airports. For instance, the DfT (2006) has acknowledged that mandatory emissions limits apply to new machinery and surface vehicles used at UK airports; such standards are also applied in many other countries. For airport terminal applications, opportunities exist to install energy-efficient, low-emission equipment; to invest in new infrastructure that uses sources of renewable energy in preference to fossil fuels; and to convert existing infrastructure to use renewable energy sources.

Fuel Options

As previously mentioned (Chapter 3), the air transport system is critically dependent upon the availability of reliable sources of kerosene and that dependence will persist over at least decadal timescale (Allen 1999; RCEP 2002). In 1999, the IPCC (1999, 10–11) stated that there 'would not appear to be any practical alternatives to kerosene-based fuels for commercial jet aircraft for the next several decades.' Progress has been made in reducing the sulphur content of kerosene which, in turn, has reduced aviation emissions of SO_x and sulphate particles. However, because reducing the sulphur content of kerosene also affects the lubricity of the fuel, the extent to which fuel sulphur content can be reduced is limited (IPCC 1999). In the long-term, though, the air transport industry, in common with other transportation sectors, faces the problem of crude oil replacement during the twenty-first century and crude oil scarcity – or high oil prices – could potentially curtail the growth of the air transport industry by 2050 (Allen 1999). Therefore, considerable efforts are being made to develop new aviation fuels, and the potential to reduce the impacts of aircraft on climate through the use of alternative fuels to kerosene has received scrutiny (Bows et al. 2009; RCEP 2002). The availability of alternative fuels, such as liquid hydrogen (H_2), biofuels, synthetic fuels and liquefied natural gas (LNG), or the use of alternative power sources, such as fuel cells, could potentially reduce the emissions generated by aircraft, although many of those innovations would require new aircraft and airport infrastructure designs (ACARE 2004).

As noted in Chapter 3, biofuels have already been developed for aviation use and some experimental trials have been undertaken; however, the production of biofuels may have other, negative impacts on progress towards sustainable development (such as influences on food prices and availability) and it is important to emphasise that biofuels are being investigated as a potential response to the challenge of climate change rather than because they offer particular benefits for air quality (Marsh 2008; Mathiesen et al. 2008; Simões and Schaeffer 2005; Takeshita and Yamaji 2008; Upham et al. 2009). Recent research has investigated the potential benefits of a proposed fleet of aircraft fuelled by liquid H_2; the combustion of liquid H_2 would emit very few particles and less NO_x than the use of conventional kerosene, although further research into this topic is required (Gauss et al. 2003; Janić 2008; Lee 2004; 2009; Marquart et al. 2003; Ponater et al. 2006; RCEP 2002). A further consideration is that the fuel options mentioned here could potentially offer substantial benefits in terms of overall emissions reductions, but their benefits over the LTO cycle (the part of the aircraft operational cycle which is of concern in relation to air quality) would be much more modest. Overall, improvements in aircraft fuel technology – like engine and airframe technology improvements – require substantial investment and are likely to be viable only in the long term. Consequently, limited potential exists to make radical progress in reducing emissions due to improved fuel technology in the short to medium term. Again, however, in relation to air quality, the fuel used by aircraft is only one consideration; fuels are also burned in other combustion processes at airports and there may be much greater scope to achieve emission reductions through the use of alternative fuels in surface transportation systems and in airport power, heating and cooling plants.

Operational Options

Operational options for mitigating the impact of aircraft on air quality involve configuring, loading, manoeuvring and maintaining aircraft in such a way as to maximise fuel efficiency: for instance, by reducing unnecessary aircraft weight, increasing load factors, ensuring high standards of aircraft maintenance, minimising route distances, optimising cruising speeds and levels, and minimising periods of holding (Bows et al. 2009; Drake 1974; Eyers et al. 2004; IPCC 1999; Simões and Schaeffer 2005). Accelerating the rate of fleet renewal is another method by which aircraft operators can improve their average fuel efficiency (Bows et al. 2009). As previously mentioned, maximising fuel efficiency has the side-effect of reducing many aircraft emissions – although not necessarily emissions of NO_x. Thus some aircraft emissions may have been inadvertently reduced as a side-effect of operational practices designed to conserve fuel. Nonetheless, there are many reasons why aircraft may still consume more fuel during flights than is strictly required to complete each sector; some of those reasons relate to airline procedures, whilst others concern ATM systems and procedures, including system-wide considerations such as the design of airspace. In general, system inefficiencies arise in the air transport system during all phases of flight – and during ground movements – as a result of congestion, leading to air and ground traffic constraints. However, in relation to air pollution at the Earth's surface, the only relevant operational phases of flight are those of the LTO cycle, meaning that operational improvements made at higher levels are unlikely to affect air quality. Even the practice of stacking, a feature of congested airspace in which aircraft are directed into holding patterns during descent until a landing slot becomes available, generally occurs above the atmospheric boundary layer, with the result that the emissions generated during airborne holding are largely insignificant in terms of air quality.

Holding on the ground, in contrast, when aircraft queue on taxiways or aprons due to surface congestion, has a direct impact on air quality in the vicinity of airports, and ATM improvements that increase the system efficiency of taxiing could offer significant benefits for air quality; reducing the time that aircraft spend taxiing could potentially yield fuel efficiency improvements in the range 2–6 per cent, and hence emissions reductions (IPCC 1999; Marín 2006). Overall, the IPCC (1999) has acknowledged that improvements in ATM and other operational systems and procedures could reduce aviation fuel consumption by between 8 and 16 per cent, with the large majority (6–12 per cent) of those reductions coming from ATM improvements that are anticipated to be implemented fully by 2020. Improvements in the efficiency of the ATM system will in principle mean that most specific aircraft emissions are reduced, although the rate at which improved ATM systems and procedures are introduced depends upon the creation of the necessary institutional frameworks at an international level, such as the implementation of the Single European Sky initiative (EUROCONTROL 2007; IPCC 1999). Again, however, such improvements may not necessarily translate into reductions in NO_x emissions because the relationship between fuel consumption and NO_x formation is non-linear. Another operational means of reducing most specific emissions from aircraft – and thus of reducing air pollution per aircraft movement – is to increase the load factor of a given flight by eliminating non-essential weight and by maximising the payload (IPCC 1999). In this way, maximum utility may be obtained for a given level of environmental impact. Loading aircraft efficiently involves a combination of (a) minimising the weight of the aircraft before its payload is stowed (for instance, by minimising the carriage of unusable fuel) and (b) maximising the passenger seat occupancy or the cargo load. Efficient loading of aircraft can also be increased by avoiding tankering – the practice of enplaning more fuel than a particular flight requires so that it can be used for subsequent flights. The effect of tankering is to increase aircraft mass and thus fuel consumption, as well as the specific emissions of most air pollutants (Drake 1974; Eyers et al. 2004; IPCC 1999).

Many other operational improvements could also reduce emissions and, in turn, could increase air quality in the vicinity of airports. Improved aircraft maintenance can help to ensure that high levels of fuel efficiency are maintained throughout the service life of the airframe and engines (Curran 2006). Many initiatives relating to ATM systems and procedures could potentially reduce emissions during the L1O cycle: improved communications, navigation and surveillance and air traffic management (CNS/ATM) systems; arrival management (AMAN) and departure management (DMAN) systems; continuous descent approaches (CDAs) and low power, low drag (LP/LD) approaches; and expedited climb departure procedures that allow aircraft to climb rapidly to their optimal cruising levels without encountering climb restrictions (Auerbach and Koch 2007; Dobbie and Eran-Tasker 2001; ICAO 2004). Such measures could reduce fuel burn during the LTO cycle and may thus improve air quality near the surface. Further operational efficiencies could be achieved on stand through the use of fixed electrical ground power (FEGP) in preference to kerosene-burning APUs. Overall, however, operational options to reduce the impacts of aviation on air quality are generally in their infancy; indeed, they are being investigated almost entirely as responses to climate change impacts rather than to improve air quality. Furthermore, improved operational efficiency may result in additional air traffic being attracted to airports, thereby negating any emission reductions that might have been gained (IPCC 1999). Consequently, the DTI (1996, 10) has acknowledged that 'operational measures will not offset the impact of the forecast growth in air travel'. Whilst operational options may form part of an overall strategy to mitigate the impacts of air transport on air quality, they are not a sufficient response to that challenge.

Policy Options

A range of regulatory, economic and voluntary measures to reduce the impact of air transport on air quality can be grouped together within the category of policy options (Bishop and Grayling 2003; Daley and Preston 2009; Roberts 2004). The potential improvements in aircraft and engine technology, in fuel technology and in operational systems and procedures (including the efficiency of the air traffic system) discussed above could reduce some of the impacts of aviation on air quality; however, they will not fully offset the effects of the increased emissions resulting from the projected growth in aviation (IPCC 1999). Due to the high abatement costs of the sector, and the limited potential for radical technological or operational solutions to be found in the short to medium term, success in reducing the impact of air transport on air quality therefore depends upon the formulation and implementation of effective policy (Bishop and Grayling 2003; Daley and Preston 2009). Many policy instruments are available; in particular, the IPCC (1999) has drawn attention to the following policy options which could be used to reduce aviation emissions: (a) more stringent aircraft engine emissions regulations; (b) removal of subsidies and incentives that have negative environmental consequences; (c) market-based options such as environmental levies (charges and taxes) and emissions trading; (d) voluntary agreements; (e) research programmes; and (f) substitution of aviation by other transport modes. However, not all of these options have been fully investigated or proven in relation to aviation (IPCC 1999). Furthermore, policy instruments (other than the application of ICAO engine certification standards) have been little used specifically to manage aviation impacts on air quality. Nevertheless, policy instruments have considerable potential to be used for that purpose.

Policy instruments are often grouped within the categories of regulatory, market-based and voluntary approaches; they include proposals to cap aviation emissions, to impose taxes and emissions charges, to use or remove subsidies, to issue tradable permits for aviation emissions and to encourage the use of voluntary agreements within the air transport industry (Bishop and Grayling 2003; IPCC 1999; Pastowski 2003; Roberts 2004). As previously acknowledged (Chapter 3), such proposals have individual strengths but are also problematic for a variety of reasons. Regulatory approaches face the problem that air transport is an international industry that spans national jurisdictions – and nations varying in their capacity to monitor and to enforce environmental standards. Market-based approaches must negotiate difficult issues related to the varying competitiveness of air transport service providers – including the need to internalise differing costs of pollution whilst facilitating access to international air transport markets on an equitable basis between nations. Voluntary approaches face the criticism that they are too weak to catalyse the profound behavioural change that is required to ensure that air transport is compatible with the requirements of sustainable development (Broderick 2009; Hewett and Foley 2000). Such objections have emerged in debates about the potential of various policy instruments to reduce the impact of air transport on climate change. Again, it is not clear that all of those considerations would apply, in the same way, at the local and regional scales that are of concern for air quality in the vicinity of airports – although similar issues may yet emerge. The main types of policy approach are discussed in turn below.

Regulatory approaches Regulatory approaches typically involve the imposition of standards, especially in cases where pollutants pose a risk to human health. Some aviation emissions are subject to regulatory standards as part of the ICAO engine certification process. During that process, emissions of HCs, CO, NO_x and smoke from a small number of new (or nearly new) aircraft engines, under simulated LTO cycle conditions, may not exceed specified values; thus the engine certification process represents a starting point for the management of aviation emissions (IPCC 1999). The ICAO engine certification standards are negotiated and agreed through the ICAO Committee on Aviation Environmental Protection (CAEP) and they generally become more stringent with time as improved technologies become available. The first CAEP emissions standards (the CAEP 1 regulations) were agreed in 1981 and took effect in 1986; subsequent standards have increased the levels of stringency applied to new engines, with the result that current NO_x standards are approximately 40 per cent more stringent than the original standard (ICAO 2007; Lee 2004). Currently, of the pollutants regulated by ICAO, only the reduction of NO_x emissions remains a significant technological and environmental challenge (Lee 2004). However, reducing NO_x from aircraft engines is far from straightforward. The task of reducing NO_x emissions has been hindered by concomitant progress in creating quieter and more fuel-efficient engines, which has had the side-effect of increasing the overall pressure ratio of engines, thereby increasing specific NO_x emissions. Due to the complex relationship existing between engine overall pressure ratio and NO_x emissions, the regulatory parameter for NO_x varies with the overall pressure ratio of an engine, which complicates the regulation of NO_x emissions (Lee 2004). A further issue is that new aircraft are required to meet relatively stringent NO_x emissions standards, whilst older aircraft tend to remain in service long after the introduction of new regulations. For these reasons, taking the global fleet into consideration, the emission factor for NO_x increased from 1992 to 1999 and is expected to continue increasing until at least 2015 (Eyers et al. 2004; Lee 2004; see Chapter 2).

The Advisory Council for Aeronautics Research in Europe (ACARE) has published several environmental goals for the air transport industry, including a reduction of 80 per cent in NO_x emissions for new aircraft in 2020 (compared with new aircraft in 2000). Such a reduction in specific emissions of NO_x would be equivalent to approximately a 60 per cent reduction in the emission factor for NO_x (ACARE 2004). Whilst those goals are not in themselves regulatory standards, it is likely that the stringency of the ICAO regulatory standards would increase with technological developments and with the spread of innovation within the air transport industry. Indeed, ICAO, through CAEP, has also set mid- and long-term NO_x technology goals which increase considerably the stringency of current certification standards: to 45 per cent below current certification levels by 2016 and to 60 per cent below current certification levels by 2026 (ICAO 2007). However, given the long lead-in times for new technologies in the aviation industry, those targets are acknowledged to be ambitious. The pace at which such technology goals are achieved and the rate at which new technologies are incorporated into the global fleet – given the high costs and long service-lives of aircraft – will affect the rate at which NO_x emissions are reduced. Again, these comments apply to NO_x emissions from aircraft; in contrast, the direct regulation of emissions from other airport sources, such as surface vehicles, may be much simpler to achieve.

Market-based approaches Market-based approaches are based on the principle of promoting environmentally desirable courses of action using economic incentives and disincentives (Chapter 3). The relevant instruments include taxes, charges, subsidies and tradable (or marketable) permits; however, very few of those instruments have been used specifically to manage aviation impacts on air quality. Emissions taxes or charges could potentially be used to reduce emissions, thereby improving air quality. However, whilst emissions are relatively easy to calculate in relation to the impact of air transport on climate, due to the linear relationship between fuel burn and CO_2 emissions, the same is not true of NO_x and particle emissions. Indeed, the difficulties involved in accurately monitoring and modelling emissions of NO_x and particles from aircraft could make such taxes or charges extremely complex and challenging to administer. Other forms of taxation – such as fuel taxation – could in principle be used to reduce NO_x and particle emissions indirectly. However, again, given the non-linear relationship between those emissions and fuel burn, such taxes and charges could be highly problematic. Moreover, the idea of fuel taxation is controversial within the aviation industry, due to the sensitivity of air transport to kerosene prices, and there are major obstacles to imposing an international tax on aviation fuel (Cairns and Newson 2006; Carlsson and Hammar 2002; ICAO 2006b; IPCC 1999; Mendes and Santos 2008; Pearce and Pearce 2000; Seidel and Rossell 2001; Stern 2007; Wit et al. 2004). Taxation could be used to restrain demand for air travel and thereby reduce emissions; such an approach could involve imposing Value Added Tax (VAT) on international air tickets (which is regarded as logistically complex) or increasing the Air Passenger Duty (APD) (which is regarded as a 'blunt instrument'; Cairns and Newson 2006; DfT 2003c). Overall, however, the prospect of using taxes and charges to reduce emissions of NO_x or particles seems remote.

Subsidies are another market-based policy instrument that could potentially be used to encourage the reduction of emissions. Subsidies may take the form of grants, loans or tax allowances. In relation to aviation impacts on air quality, subsidies could be used to accelerate fleet replacement, to promote the development and use of alternative fuels (and other technologies) and to encourage the introduction of low-emissions surface vehicles and stationary equipment at airports (Lambert 2008; RCEP 2002). However, aviation already benefits from a range of economic incentives that have allowed the industry to avoid paying the full environmental costs of its activities, and some commentators have argued that the trend in environmental policy for aviation should ideally tend towards the removal of current subsidies and privileges within the sector rather than the creation of new ones (EC 2006; Peeters et al. 2006; T&E 2006). Tradable (or marketable) permit schemes represent another market-based policy instrument; those schemes give polluters incentives to reduce pollution by creating new markets with defined property rights (Gander and Helme 1999; Seidel and Rossell 2001, 29; see Chapter 3). For the air transport industry, emissions trading has been discussed – and introduced in the EU – in relation to greenhouse gas emissions, but emissions trading schemes have not yet been developed for the purpose of managing the impacts of aviation on air quality (DfT 2004; Lee 2004). Overall, market-based approaches offer significant potential for reducing the impact of air transport on air quality, especially if they are used in conjunction with other measures. Recently, ICAO, through CAEP, has established a Market Based Measures task force to investigate the management of aviation impacts on air quality; that group is investigating market-based measures to reduce emissions of NO_x, SO_x, particles, CO and HCs, including emissions trading schemes.

Voluntary approaches Voluntary approaches to mitigating the impacts of air transport on air quality represent another type of policy instrument. Voluntary approaches depend upon organisations and individuals making decisions to reduce their emissions of key air pollutants, even in the absence of regulatory requirements or economic incentives to do so. Such decisions may be motivated by a variety of concerns including the belief that working co-operatively with regulators may lead to more sympathetic regulation of their operations. Polluters may perceive greater opportunities to influence the regulatory process if they can demonstrate substantial voluntary efforts to improve their environmental performance. Polluters may also initiate voluntary emission reduction measures in anticipation of stricter regulation in the future, especially if early adoption offers them a competitive advantage; they may voluntarily adopt cleaner processes in order to standardise their operations across countries or regions; and they may seek to maximise their access to worldwide markets by adopting processes that would comply with the environmental regulations of the strictest country. Ultimately, organisations may voluntarily improve their environmental performance if they believe that consumer expectations require such action. Hence companies can improve their consumer relations and brand images by demonstrating corporate responsibility, either environmentally or socially (Broderick 2009). In relation to the impact of air transport on air quality, however, voluntary approaches have been little used (although air quality side-benefits may have resulted from other voluntary initiatives, such as voluntary offsetting schemes for CO_2 emissions, or from the ongoing initiatives to improve fuel efficiency that are commonplace within the air transport industry). Hence there exists significant potential for voluntary measures to be undertaken by airframe and engine manufacturers, airlines, airports and ATM service providers specifically to reduce the impact of air transport on air quality, including the introduction and promotion of a wide range of corporate responsibility initiatives.

Summary

Air transport has a relatively small but nonetheless significant – and increasing – effect on air quality (Peace et al. 2006). Whilst many of the pollutants emitted by aircraft and other airport sources are relatively insignificant in terms of their effects on air quality, some emissions, especially of NO_x and particles, are of critical concern since they may (and sometimes do) cause air quality objectives to be exceeded, especially in the vicinity of major airports (DfT 2006; Unal et al. 2005). The impacts of those key pollutants on air quality are likely to be compounded both by the projected growth in demand for air transport and by the associated increase in demand for surface transport that would accompany further aviation growth. At the same time, regulatory standards and air quality objectives are becoming increasingly stringent as public tolerance of air pollution decreases. In particular, concerns about NO_x and particle emissions from aircraft and from other airport sources can potentially constrain airport infrastructure development, especially at large airports, and may thereby restrict aviation growth more generally. Indeed, several European airports – including Zurich, Amsterdam Schiphol and London Heathrow Airports – are already subject to stringent air quality constraints. With the introduction in the EU, in 2010, of mandatory air quality limits for NO_2 – including an annual mean limit of $40\mu g \; m^{-3}$ – together with the long-term goals set by ICAO and ACARE to reduce NO_x emissions, the issue of aviation impacts on air quality is becoming increasingly acute. The fact that regulatory standards are tightening whilst NO_x emissions from international aviation are increasing rapidly (almost doubling in the EU during the period 1990–2005) presents policymakers with a conundrum (EEA 2006).

Whilst some of the effects of air transport on air quality are relatively well–understood, scientific understanding of some aviation pollutants and the processes by which they are generated and transported remains comparatively limited. Consequently, research programmes have been established in order to develop better understanding of the formation and evolution of pollutants within the engine and in the exhaust plume, to provide improved knowledge of the dispersal of aviation emissions and to create a clearer picture of the impacts of key pollutants – especially of particles – on human health. Further research is also required to investigate the likely effectiveness of various potential emission reduction scenarios. Whilst many issues remain to be clarified by research, it is clear that it is becoming increasingly difficult to find technological solutions to the problem of aircraft emissions. Ongoing work focuses on the design of aircraft engines – particularly of fuel combustors – in order to achieve specific NO_x reductions whilst simultaneously improving aircraft fuel efficiency and noise performance. Attention is also focused on other technological measures, such as the use of alternative fuels, and on operational measures to reduce emissions. Continuing, incremental advances in reducing specific aircraft emissions are expected to be made, although those improvements are being vastly outpaced by the growth of air transport, meaning that the absolute emissions of some air pollutants – including NO_x – are rapidly increasing.

Once again, a sustainable development dilemma confronts policymakers: the need to address the impact of aircraft and airport operations on air quality (and simultaneously on other parts of the environment) without blighting the economies and societies that depend upon air transport (see Chapter 6). Therefore, the need to devise and implement an effective policy response to the impact of air transport on air quality is becoming critical, especially at the local scale in the vicinity of major airports. A variety of options is available to policymakers in that task, including the use of regulations, emissions taxes and charges, subsidies (or their removal), emissions trading schemes and voluntary measures. Some of those policy instruments have considerable potential because, to date, they have been little used to manage the impact of air transport on air quality. The ICAO aircraft engine certification standards are applied at the point of engine manufacture, but those standards represent the only regulatory control of aviation air pollution and aviation emissions are not directly regulated at most airports. In the short to medium term, policy approaches are likely to focus on encouraging greater use of the most fuel-efficient (and least-polluting) aircraft, especially at major airports where compliance with air quality objectives is already marginal. In the longer term, a combination of policy approaches may be adopted: the use of voluntary agreements within the industry to reduce emissions; the imposition of more stringent emissions standards, especially for NO_x and particles; the removal of existing privileges and subsidies of the industry; the wider use of emissions charges, fuel taxes and other levies; the development of emissions trading schemes for key air pollutants; and, ultimately, the use of demand restraint measures and the restriction of airport infrastructure development. Given the unpopularity of such measures, in the long term, the development of the air transport industry depends critically on finding radical technological solutions to reduce the NO_x and particle emissions associated with flight. Such a conclusion is similar to that reached for climate change (see Chapter 3), where CO_2 emissions are the dominant – although not the only – issue of concern. A confounding factor for policymakers, however, is that tackling one of these environmental issues (CO_2 emissions) hinders progress towards solving the other (NO_x emissions) and *vice versa*. Thus the 'technological fix' sought by the air transport industry – if current growth trends are to continue – must somehow reconcile reductions in CO_2 emissions with concomitant reductions in NO_x and particle emissions.

5   Aircraft Noise

Introduction

Aircraft noise is acknowledged to be one of the most significant local environmental impacts associated with the operation of airports (DfT 2002; 2003b; 2007; FAA 2005; Hume and Watson 2003; Hume et al. 2003; Kryter 1967; May and Hill 2006; Nero and Black 2000; Pattarini 1967; Skogö 2001; Thomas and Lever 2003). The nuisance caused to individuals by aircraft noise is an important issue affecting communities in the vicinity of airports and their flight paths. Given that major airports are often located within, or in close proximity to, urban areas, aircraft noise affects large numbers of people including many who do not benefit directly from the provision of air transport services (Hume and Watson 2003). It has long been recognised that the issue of aircraft noise nuisance is not straightforward but involves the complex interaction of many factors, including a range of physical, physiological, psychological and sociological processes (Schultz 1978). Yet despite the complexity of the issue, aircraft noise represents a common, significant source of annoyance that can affect many aspects of people's lives: by interrupting communication and leisure activities, by disrupting activities requiring concentration and by discouraging people from using outdoor spaces. Aircraft noise is also responsible for disturbing many people's sleep. Furthermore, the experience of aircraft noise may exacerbate conditions of stress, anxiety and ill-health for many people, especially those within more vulnerable social groups, such as children, elderly people and individuals with pre-existing illnesses. As a result of those wide-ranging effects, the impact of aircraft noise can severely reduce the wellbeing of individuals, especially in the vicinity of major airports and their flight paths (Hume and Watson 2003).

To some commentators, the persistence of aircraft noise as an important environmental issue is surprising given the major advances in airframe and engine design that have been achieved since the 1960s (Thomas and Lever 2003). Those improvements have been driven, in part, by the introduction of increasingly stringent noise regulations by the International Civil Aviation Organization (ICAO), which have prompted both the development of quieter aircraft and the retirement from service of older, noisier variants. The distribution of those improvements in aircraft noise performance has not been uniform; the most dramatic advances in aircraft noise management have occurred in Europe, North America and some parts of Asia, reflecting the greater prominence of environmentalist concerns in those regions, whilst noisier aircraft continue to operate in other parts of the world. Overall, however, the trend has been towards the introduction and increased use of quieter aircraft, with some airports being available only to the operators of the quietest aircraft in a given class, especially at night. Consequently, for a certain period since the 1980s, the number of people exposed to high levels of aircraft noise around many major airports declined (Thomas and Lever 2003). Yet the effective management of aircraft noise is confounded by several factors. Most importantly, whilst incremental improvements in aircraft noise performance are still being made, those advances are now being offset by the dramatic growth of air traffic (Hume and Watson 2003; see Chapter 2). Therefore, in general, aircraft noise exposure around many airports is now increasing. Moreover, much of the annoyance reported by noise-affected people may now be due as much to the frequency with which people are over-flown as to the sound levels generated by individual aircraft movements (Smith 1992; Thomas and Lever 2003). A further consideration is that, as average affluence levels rise in many countries, particularly in industrialised societies, individuals living near to airports and their flight paths are becoming increasingly sensitive to environmental issues such as noise nuisance. In many places, people living near airports have increasing expectations with respect to their health, wellbeing and quality of life, together with decreasing tolerance of noise and other environmental impacts (Hume and Watson 2003; Moss et al. 1997). The problem of aircraft noise nuisance is thus a complex, subjective and political topic that involves many issues of perception, representation, tolerance, affluence, power and justice (Maris et al. 2007; May and Hill 2006; Thomas and Lever 2003).

The task of understanding and managing the issue of aircraft noise more effectively is a critical one for the air transport industry because increasing levels of noise nuisance – together with decreasing tolerance of that noise nuisance – can generate fierce community opposition to airport operations and infrastructure development proposals. Thus public concerns about aircraft noise represent a major constraint to the growth of the air transport industry. As a result of such profound concerns about aircraft noise, various noise-related operational restrictions have already significantly constrained the growth of some major airports, particularly in Europe (EUROCONTROL 2004; 2008; ICAO 1993). In many places, the task of managing airport development (and aviation growth more generally) now depends upon the extent to which policymakers succeed in reconciling the desire for aviation growth with other important considerations: the need to protect people – especially vulnerable social groups – from the unacceptable impacts of aircraft noise; the need to address a wide range of other airport-related community concerns; and the need to manage other aviation environmental impacts effectively (see Chapters 3 and 4). In that context, this chapter explores the various issues associated with aircraft noise, including the development of scientific knowledge and understanding of aircraft noise and its impacts. The account presented below begins with a brief discussion of relevant terminology and measures, and of some common ways in which aircraft noise is measured and reported. Next, the impact of aircraft noise is considered; that impact includes a range of effects including annoyance (due to the interruption of communication and leisure activities, the disruption of activities requiring concentration and the discouragement of outdoor activities) and sleep disturbance. However, such categories are not always clear-cut and considerable overlap may occur between different noise-related impacts. The fact that aircraft noise may have a significant impact on human wellbeing and quality of life, especially for more vulnerable individuals, underlines the importance of reducing those impacts as far as possible – as does the fact that aircraft noise nuisance prompts many complaints to airport operators, and considerable opposition to airport infrastructure development, from members of the public. This chapter provides a discussion of the main ways in which the impact of aircraft noise may be reduced, including various technological, operational and policy options. In that task, trade-offs must inevitably be made between the management of aircraft noise and other environmental issues such as the impact of air transport on climate change (see Chapter 3).

Aircraft Noise

The phenomenon of aircraft noise involves the generation, propagation, sensation and perception of the sound associated with aircraft movements. In physical terms, that sound may be characterised in terms of variations in air pressure, wavelength, frequency, amplitude and purity (Sekuler and Blake 1994; Thomas and Lever 2003; Veitch and Arkkelin 1995). A commonly-used measure of sound intensity is the decibel (dB) which expresses the magnitude or power of a sound using a logarithmic scale in relation to the average human threshold of perception of sound. A related scale of measurement is the A-weighted decibel scale, written as dB(A), which gives a weighting to certain sound frequencies according to the sensitivity range of the human ear, reflecting the fact that the human ear does not detect all frequencies of sound with equal efficiency. Thus the A-weighted decibel scale most closely approximates the response of the human ear to sound. However, whilst such terms may be used to describe the physical response of the human ear to sound, they do not measure the total quality of the sensation perceived by a subject – or the degree of nuisance caused to an individual as a result of hearing that sound. The FAA (1985, 1) has stated that sound 'is a complex vibration transmitted through the air which, upon reaching our ears, may be perceived as beautiful, desirable, or unwanted. It is this unwanted sound which people normally refer to as noise.' Thus 'noise' may be defined as a loud, discordant or disagreeable sound (or sounds); noise is often regarded as a sound that has an undesirable effect upon people. 'Ambient noise' may be defined as the constant, spontaneously occurring background noise which is the sum of the individual noises at a given location, including noises of varying levels generated at varying distances from the observer. Airports are the sites of many noise-generating processes including the operation of aircraft, maintenance machinery, heating and cooling systems, power plants and ground vehicles; hence the ambient noise levels at airports may be relatively high. However, with the exception of the noise generated by aircraft, the ambient noise levels at airports may be comparable with that generated at other large commercial or industrial sites. Therefore, in relation to air transport, public concerns about the effects of noise focus predominantly on the noise generated by aircraft. Importantly, aircraft noise is not created solely by aircraft engines but also results from the movement of turbulent air over the airframe – especially when the aircraft is operated in a high-drag configuration, such as with the landing gear and flaps lowered or with the speedbrakes deployed (Smith 1992). Considerable noise may also be generated by aircraft on the ground during runway operations, taxiing, engine testing, and the use of auxiliary power units (APUs).

As the following section explains, the experience of aircraft noise is highly subjective and it affects people in complex, interrelated ways. As a result, the ideal measure of aircraft noise would express the total nuisance or annoyance caused to each individual as a result of aircraft operations. In practice, however, such a measure would be impossible to devise and to interpret. Due to the difficulties involved in precisely determining individual responses to aircraft noise, it is necessary to adopt a more general approach to characterising overall noise exposure – an approach that involves quantifying and averaging noise exposure.

The level of exposure to aircraft noise may be measured in a variety of ways (Table 5.1). As yet, there is no single, standard measure to describe the noise environment surrounding airports, despite recent efforts to harmonise practice (Europe-wide, for example, in EC Directive 2002/49/EC on the assessment and management of environmental noise; EC 2002). Various types of measure are used to measure aircraft noise: measures of the sound associated with individual aircraft movements (single event measures); measures of the total sound generated by the operation of many aircraft over a given period of time (cumulative scores); and measures of the average sound level created by many aircraft during a given period of time (time-averaged measures). Single events are often described in terms of the instantaneous maximum sound level (L*max*) or the sound exposure level (SEL), both of which indicate the maximum instantaneous (or very short-term) sound level experienced at specified points on the ground during an aircraft overflight. In contrast, time-averaged measurements describe the average energy associated with aircraft noise over a particular time period, and they may be differentiated into day-time, evening and night-time periods. For instance, the equivalent continuous sound level (L*eq*) measure is commonly used as a long-term average noise exposure measure; a 16-hour L*eq* value of 57dB(A) indicates that a particular (fluctuating or intermittent) noise source has generated a sound energy level equivalent to that of a constant sound of 57dB(A) over a 16-hour period. Time-averaged measures are frequently used to characterise aircraft noise because they illustrate the average sound energy generated in the vicinity of airports, so may be used to create maps of noise exposure. However, many commentators have expressed misgivings about the use of time-averaged noise measures since – as with any averaging process – they conceal considerable variation between individuals and places, and they also fail to reflect the extreme nuisance caused by disproportionately loud aircraft movements. Some commentators have argued that aircraft noise nuisance may be more closely related to extreme, single noise events than to cumulative energy measures – an issue that is discussed in more detail below (Hume et al. 2003).

The most commonly-used approaches to measuring aircraft noise exposure are based on time-averaged measures; they involve the calculation of the average sound energy delivered – which is related to the extent of noise exposure – at a particular location over a specific time period (typically eight, twelve or sixteen hours). Therefore, average sound energy levels are a function of both the number of aircraft movements and the sound generated by each movement. Strictly, the time interval should be specified by a subscript (such as L*eq*$_8$ for an 8-hour noise exposure

Table 5.1   Some measures of aircraft noise

Measure	Meaning
SINGLE EVENT MEASURES	
Instantaneous maximum sound level ($Lmax$)	The maximum (A-weighted) sound level recorded at specified points on the ground during a single aircraft overflight; however, $Lmax$ does not provide any information about the duration of that sound
Sound exposure level (SEL)	The total sound energy recorded at specified points on the ground during a single aircraft overflight; SEL provides more information than $Lmax$ as it includes information about both the duration and the magnitude of that sound; SEL is not an instantaneous measure of aircraft noise but refers to a short period (seconds) during an overflight
TIME-AVERAGED MEASURES	
Equivalent continuous sound level (Leq)	The average sound energy recorded over a period of time (multiple overflights); a 16-hour Leq value of 57dB(A) indicates that overflights have generated a sound energy level equivalent to that of a constant sound of 57dB(A) averaged over a 16-hour period; however, Leq does not include any evening or night-time weighting
Day-night average sound level (DNL or Ldn)	The average sound energy recorded over a 24-hour period; it includes a weighting to reflect increased human sensitivity to noise at night; a weighting of 10dB is added to the night-time (2200-0700 hours) sound levels
Day-evening-night average sound level ($Lden$)	The average sound energy recorded over a 24-hour period; it includes weightings to reflect increased human sensitivity to noise in the evening and at night; weightings of 5dB and 10dB are added to the evening (1900-2300 hours) and night-time (2300-0700 hours) sound levels, respectively
Community noise equivalent level (CNEL)	CNEL is similar to $Lden$; it describes the average sound energy recorded over a 24-hour period, with weightings of 5dB and 10dB added to the evening (1900-2200 hours) and night-time (2200-0700 hours) sound levels, respectively
CUMULATIVE SCORES:	
N70 contours	N70 contours join places in which a certain number of individual noise events in a given period exceeds 70dB(A) $Lmax$

period) unless it is clearly understood from the context; in common usage, L*eq* values tend to refer to the 16-hour 'daytime' period. One notable consideration is that L*eq* does not include any form of evening or night-time weighting; the sound is assumed to be equivalent whenever it is produced. Another measure of continuous noise exposure is the day-night average sound level (DNL) measure, also written L*dn*. Like L*eq*, the L*dn* measure describes the 24-hour average sound level, ideally for an average day over the course of a year; however, it includes a weighting to reflect the generally increased human sensitivity to noise nuisance during the hours of night, when ambient noise levels are lower and when many people are trying to sleep. The weighting is achieved by the addition of 10dB to night-time sound levels (defined as 2200–0700 hours). A further refinement is found in the day-evening-night average sound level (L*den*) measure, which is weighted to take account both of the additional nuisance caused by noise in the evening (1900–2300 hours with a weighting of 5dB) and during the night (2300–0700 hours with a weighting of 10dB). The community noise equivalent level (CNEL) measure is similar to L*den*; it also describes the average A-weighted noise exposure over a (typically) 24-hour period, with a 5dB weighting for the evening period (1900–2200 hours) and a 10dB weighting for the night-time period (2200–0700 hours). In contrast to those measures of average sound exposure, the single event measures focus on individual occurrences of noise, usually above a specified minimum threshold, caused by an intrusive noise source such as an aircraft overflight. L*max* is a commonly-used single event measure which describes the maximum A-weighted sound level for a given noise event; thus L*max* provides information about the instantaneous maximum sound level recorded during the overflight of an aircraft – but not about the duration of that sound. The SEL measure – also a single event measure – provides more information than L*max*; SEL measures the total sound energy associated with an individual noise event, including information about both the duration and the magnitude of the sound. Strictly speaking, SEL is not an instantaneous measure of aircraft noise but instead describes the sound over a very short period (seconds) as an aircraft passes a particular point. Other measures of noise exposure also exist, including measures of sound levels occurring specifically in the evening (L*evening*) and at night (L*night*).

All noise measures, however, require careful interpretation. One typical use of noise measurements is to provide visual representations of the sound environment in the vicinity of an airport using noise exposure contours. Noise exposure contours take the form of a series of closed curves plotted on a map, with each contour joining locations receiving the same average sound energy due to aircraft movements; thus noise contours are analogous to those linking places of the same elevation on a topographical map. One commonly-used method of portraying the sound environment in the vicinity of an airport is to produce L*eq* noise exposure contours, which join places with equal L*eq* values (for instance, the 57dB(A) L*eq* contour); those values include the sound generated at a given airport by both arriving and departing aircraft. Since, at some airports, aircraft noise occurs predominantly during the day, the L*eq* values (which reflect average sound levels) for those airports could be misleading, due to the low levels of noise during the night, and this could distort the representation of day-time aircraft noise exposure (Greaves and Collins 2007). For this reason, L*eq* noise exposure contours are generally produced for the day-time period (0700–2300 hours) and the 57dB(A) L*eq* area refers to the area enclosed within that contour. Noise exposure contours may also be produced for the SEL measure; SEL contours illustrate the distribution of noise energy generated from a single aircraft movement (for example, a take-off, landing or overflight), with the contours linking places with equal SEL values. Thus a 90dB(A) SEL 'footprint' illustrates the area in which SEL values equal or exceed 90dB(A). Another type of noise exposure contour is the N70 contour, which is based on the use of the L*max* measure. N70 contours are produced in order to illustrate the distribution of places where a certain number of individual noise events exceeds 70dB(A) L*max*. Typically, a series of N70 contours is plotted in order to identify areas in which varying numbers of noise events – perhaps ranging from 10 to 500 noise events exceeding 70dB(A) L*max* – are experienced. One factor taken into consideration in producing noise exposure contours is their differing complexity and ease of production. In practice, SEL contours are rarely generated; they require noise levels to be calculated for each individual aircraft type on every route, so they are much more labour-intensive to produce than L*eq* contours, which can be generated relatively easily using standard aircraft noise modelling software such as the Federal Aviation Administration (FAA) Integrated Noise Model (INM; see Arafa et al. 2007).

Noise exposure contours and footprints are potentially useful insofar as they can give a visual impression of the total noise environment of an airport, and they may also be used to illustrate the relative contributions of various aircraft types, routes and operating procedures to the overall noise exposure in the vicinity of an airport. In particular, if suitable measures are used, then noise exposure contours and footprints may illustrate the noise exposure due to aircraft movements at night – which may, in turn, be used to assess the relationship between aircraft noise and sleep disturbance.

Noise contours can also be used to identify those areas in which airport operators are required to implement compulsory purchase or sound insulation schemes, since they delineate those areas in which individuals may be exposed to levels of aircraft noise that are defined as unacceptable. In conjunction with sophisticated noise modelling techniques, noise exposure contours may be used to evaluate options for the development of new airport infrastructure, new aircraft technologies and revised operating procedures; for example, the effects of constructing a new runway on the noise environment of an airport may be simulated using advanced noise models. Calculations may then be performed to reveal any changes in the size of the noise exposure contours and – given sufficient input data – any changes in the size of the population exposed to aircraft noise. However, whilst noise exposure contours may have certain uses, they do not provide any information about how aircraft noise affects people's lives – or about whether individuals regard those effects as being acceptable.

To investigate such issues, noise exposure values, noise exposure contours and noise footprints may be compared with the results of social surveys of perceived levels of nuisance in areas of different noise exposure. Thus, for instance, the pattern of noise-related complaints by local residents to an airport operator may be compared with the distribution of average noise exposure (using Leq or DNL contours) and with single event levels (using $Lmax$ or SEL values). In general, the pattern of noise complaints received by airport operators suggests that, as traffic has increased and as individual aircraft movements have become quieter, the frequency of exposure to aircraft noise has become an increasingly important cause of nuisance (DfT 2007; Janić 1999). In addition, the pattern of noise complaints is strongly influenced by the timing of aircraft noise, especially at night, and by the occurrence of particularly loud overflights (Hume et al. 2003). Hence the relationship between noise exposure and perceived nuisance is not straightforward but is influenced by other factors, with the implication that conventional noise assessment measures (such as Leq) may, if used alone, inadequately represent the true scale of noise nuisance in the vicinity of an airport. Increasingly, innovative methods of communicating both the frequency and intensity of aircraft noise (perhaps using N70 contours or using combinations of visual and aural techniques) are required (Berckmans et al. 2008). Recent research into the communication of aircraft noise impacts to stakeholders has shown that the use of a broad range of techniques may best help individuals to understand the noise environment of an airport. In particular, the use of Leq-type measures has generated some concerns about their suitability for communicating information about noise exposure to those communities affected by aircraft noise (Thomas and Lever 2003). Information based on the use of L*eq*-type measures may be useful for comparing average sound exposure levels at different airports, or under different operating regimes at the same airport, and hence L*eq* and DNL represent (at least) a standard tool for use in policymaking (Revoredo and Slama 2008). However, L*eq* and DNL values may be difficult for lay audiences to interpret and they should ideally be supplemented with additional information – such as N70 contours – to represent aircraft noise exposure in a more accessible form (DOTARS 2000).

The use of noise exposure contours – and the various ways in which aircraft noise data have been contested and debated – raises a key question: which values of those various measures are the most significant ones for describing and managing the impact of aircraft noise? Unsurprisingly, that question has generated fierce (and ongoing) controversy (DfT 2007). A sound level of 1dB(A) is considered to be the quietest sound detectable by humans; and, since the decibel scale is logarithmic, an increase of 3dB represents approximately a doubling of the sound level. Conventionally, the 57dB(A) L*eq* value has been accepted – for instance, by the UK Department for Transport (DfT) – as the sound level that represents the onset of 'significant community annoyance' due to aircraft noise, and the 90dB(A) SEL level (as measured outdoors) has been regarded as the sound level required to awaken the average person from sleep (DfT 1992; 2003b; 2007). The significance of the 57dB(A) L*eq* value emerged following the publication of the results of the United Kingdom Aircraft Noise Index Study (ANIS), a major survey of attitudes to aircraft noise in the UK, which was undertaken in 1982 and published in 1985 (DfT 2007). Other commonly-cited 'standard' values include 50dB(A) L*eq*, below which level noise-related annoyance is regarded as being unlikely; 55dB(A) L*eq*, at which level aircraft noise is acknowledged to cause severe annoyance to some people; and 90-100dB(A) SEL, at which level the chance of the average person being awakened by an aircraft noise event is thought to be approximately 1 in 75. Needless to say, all of those values are average and probabilistic. In any community, because human responses to environmental noise are highly subjective, some people are unaffected by relatively loud noises whilst others are severely affected at lower sound levels. Nonetheless, the use of such 'standard' values has become commonplace and has important implications for policy (DfT 1992). Yet those values may be subject to critical review; for instance, research undertaken in 2005 as part of the Attitudes to Noise from Aviation Sources in England (ANASE) study has suggested that, due to aviation growth and changing public attitudes, significant annoyance may now be caused by aircraft noise levels much lower than 57dB(A) L*eq* (DfT 2007).

The highly subjective nature of human responses to environmental noise makes the task of quantifying the relationship between aircraft noise and annoyance a difficult one. Nevertheless, using such 'standard' values, it is possible to estimate various effects of aircraft noise such as the number of awakenings resulting from a given aircraft movement. That calculation involves combining information about the size of the population within the 90dB(A) SEL contour, the number of aircraft movements exceeding 90dB(A) SEL during a given time period and the probability of the average person being awakened from sleep. Such an approach, whilst mechanistic, underlines the point that individuals tend to be awakened by single noise events (as measured using the L*max* or SEL measures) rather than by high average noise levels (as measured using the L*eq* or CNEL measures). Inevitably, defining the sound levels at which sleeping individuals are awakened, or at which people who are already awake experience significant annoyance, is a contentious – and political – process. A vast range of other factors influences the subjective experience of aircraft noise, making the use of 'standard' values – such as 57dB(A) L*eq* or 90dB(A) SEL – highly problematic. In particular, aircraft and engine types, frequencies of overflight, the economic and social circumstances of populations and public levels of environmentalist concern and tolerance of noise nuisance have all changed since the 57dB(A) L*eq* value was first selected to represent the onset of significant community annoyance in the UK, in 1985 (DfT 2007). As a result, the use of conventional measures of aircraft noise now results in relatively poor predictions of the actual numbers and locations of complaints about aircraft noise – particularly of night noise nuisance and sleep disturbance. Conventional noise measures are also relatively poor predictors of the impact of aircraft noise on human wellbeing. Consequently, further research is required to determine the effectiveness of conventional noise measures in describing the actual effects of aircraft noise – and to develop better measures of aircraft noise and its impact on people (Caves 2003; DfT 2007). That impact forms the subject of the next section.

The Impact of Aircraft Noise

As the DfT (2002) has acknowledged, aircraft noise is 'widely recognised to be one of the most objectionable impacts of airport development and an important environmental issue for those living close to airports as well as further afield under the main arrival and departure tracks'. Aircraft noise has a variety of effects on people. Many of those effects are complex, subjective and interrelated; they may be categorised broadly as annoyance, sleep disturbance and other effects on wellbeing and quality of life. The first of those categories – annoyance – covers a wide range of effects such as the interruption of communication and leisure activities, the disruption of activities requiring concentration and the effect of discouraging people to use outdoor spaces. Whilst the noise generated by commercial civil aircraft is not sufficiently loud as to cause direct hearing loss amongst members of the public, chronic exposure to aircraft noise may cause people to feel stressed and angry and it may affect people's psychological and physiological functioning, their motivation and their cognitive processes (Evans and Lepore 1993; Thomas and Lever 2003; Veitch and Arkkelin 1995). Aircraft noise may also cause feelings of frustration and powerlessness amongst those who suffer its effects. Overall, such effects may amount to a significant adverse effect on people's wellbeing and quality of life. Yet, despite the number of complaints that are made about aircraft noise, many uncertainties remain about the precise nature of its effects. In particular, despite extensive research into the effects of aircraft noise on human health, the results are often contradictory, ambiguous or inconclusive and the causal mechanisms remain largely unknown. Indeed, defining 'health' is itself far from straightforward. The World Health Organization (WHO) has stated that health is a state of complete physical, mental and social wellbeing, not merely the absence of disease and infirmity (WHO 1999). However, if the definition of health is broadened to include the concept of wellbeing, then debates about the effects of aircraft noise also expand to include a range of more general problems: stress, anger, anxiety, frustration, sleep disturbance, impaired cognition and performance, interruption of communication, disruption of leisure activities, altered mood and depression (Thomas and Lever 2003).

It has long been understood that the problem of aircraft noise nuisance involves the complex interaction of many physical, physiological, psychological and sociological processes (Schultz 1978). Physical factors include those that influence the generation of sound: the aircraft and engine type, aircraft configuration, mode of operation and ambient meteorological conditions; those factors affect the resulting amplitudes, frequencies and other qualities of sound (Hume et al. 2003). In addition to those physical factors are other, human factors, including variations in the physiological systems concerned with audition, and a wide range of psychological and sociological factors – such as general health, stress, anxiety, beliefs, values, expectations, socio-economic status, cultural background and lifestyle – that affect the ways in which those auditory sensations are interpreted (Job 1996; Thomas and Lever 2003). To complicate matters further, whilst some individuals complain about the 'noise' generated by aircraft, their complaints may actually reflect various other concerns such as the fear of air accidents or the nuisance caused by other airport-related activities (such as congestion due to road traffic in the vicinity of the airport). The sound of an aircraft may therefore act as a trigger for noise-related complaints, although further investigations may reveal that the underlying cause of annoyance is not necessarily the noise but rather another nuisance associated with the operation of the airport (Job 1996; Moss et al. 1997; Thomas and Lever 2003). Some typical complaints about aircraft 'noise' in fact relate to the following subjects: (a) frustration about congestion and obstructions due to road traffic in the vicinity of airports, and especially due to airport-users parking their cars in residential areas; (b) objections to the smell of unburned hydrocarbons (HCs) arising from the evaporative emissions of aviation fuel, as well as more general concerns about the effects of air pollution on health; (c) fear of air accidents; (d) financial concerns about the loss of value of, or the inability to sell, property in the vicinity of an airport; and (e) anxieties about the effects of airport growth and infrastructure development.

Therefore, the level of perceived nuisance due to aircraft noise is only partly determined by the frequency and noisiness of aircraft movements. Perceptions of aircraft noise nuisance are also affected by the following factors: (a) affluence, attitudes, beliefs, culture, lifestyle and values; (b) awareness and acceptance of the economic and social benefits of air transport and of the potential consequences for local, regional and national economies and societies of constraints to aviation growth; (c) socioeconomic status, local transportation availability and needs, and dependence upon air route development; and (d) extent and power of public debate about, and opposition to, airport development, together with the extent to which policymakers are responsive to such debate or opposition (Thomas and Lever 2003). Aircraft noise affects individuals in different ways and can result in people making significant compensatory changes to their lifestyles. Thomas and Lever (2003) have acknowledged that the actual effects of aircraft noise vary between individuals and between airports depending upon local conditions. For example, in places with warmer climates where many people spend more time outdoors, aircraft noise is more intrusive than in colder places where people may spend much of their time inside well-insulated buildings. In many places, seasonal variations in aircraft noise nuisance are also apparent, both because air traffic levels tend to be higher in summer than in winter and because many people prefer to pursue outdoor leisure activities and to open windows in summer. Nevertheless, there are some recurrent themes in human responses to aircraft noise. Aircraft noise can prevent people from falling asleep; it can cause awakenings during the night; and it can disturb people's leisure activities, communications, socialising and studying. Some places are associated with particular sensitivity to noise nuisance, including hospitals, libraries, schools, national parks and nature reserves. In such places, aircraft noise may disrupt education and training, may cause anxiety to patients, may disturb the behaviour of wildlife and may destroy the tranquillity often associated with natural habitats.

Given the very high amplitudes of sounds associated with aircraft operations at high thrust settings, such as at take-off, during the use of reverse thrust during landing, during missed approaches ('go-around') and during some types of engine testing, questions have been raised about the possibility of aircraft noise having direct impacts on human hearing. However, auditory damage is unlikely to be caused by the noise of commercial civil aircraft (although such damage may occur in some cases: for instance, if airport workers have inadequate ear protection; Health Council of the Netherlands 1999). Nonetheless, aircraft noise is increasingly acknowledged to have effects on the health and wellbeing of recipients (Haralabidis et al. 2008; Health Canada 2001; Health Council of the Netherlands 1999; Hume and Watson 2003; Stansfeld et al. 2005). Human responses to aircraft noise that have been linked with possible effects on health include annoyance and anger, stress, anxiety and sleep disturbance, although those categories overlap and may be mutually reinforcing. Sleep disturbance may cause annoyance, for instance, and may become a source of stress and anxiety. In the past, research into the impacts of aircraft noise on health and wellbeing has focused on the following areas: surveys of the attitudes held by residents in communities near airports, using questionnaires and focus groups; physiological monitoring of individuals in laboratories or field settings during actual or simulated noise events; and interrogation of archived medical records of people living in noise-affected localities, who may report difficulty in sleeping due to night flights, and the comparison of those records with others for people in unaffected areas (Hume and Watson 2003). However, the relationship between exposure to aircraft noise and impacts on health and wellbeing is complex, subjective and non-linear; consequently, identifying clear exposure-response interactions is extremely difficult. As a result, some researchers have attempted to use other techniques to assess the impact of aircraft noise on health and wellbeing, such as the use of proxies – including the devaluation of house prices in noise-affected areas (Tomkins et al. 1998). Such an approach is not straightforward, however, because populations in noise-affected areas are to some extent self-selected, with 'coping survivors' living in the worst-affected areas (Hume and Watson 2003). In an alternative approach, researchers have used the records of complaints made directly to airport operators in an attempt to understand those concerns and objections and to identify the underlying behaviour patterns of those who complain (Hume et al. 2003). That approach was proposed by a WHO task group, which identified the need for longitudinal studies to investigate the causal relationships between mental health effects, annoyance and spontaneous complaints to airport operators (Berry and Jiggins 1999). Yet there are many confounding factors – such as the availability of personal benefits due to the presence of an airport (such as employment or convenience) – that may conceal or modify negative attitudes towards aircraft noise in such studies (Flindell and Stallen 1999). The main effects of aircraft noise – annoyance, sleep disturbance and other effects on wellbeing and quality of life – are discussed in turn below.

Annoyance

Annoyance – which may include stress and anger – is a subjective reaction frequently reported in response to aircraft noise (DfT 2007). The reported level of annoyance is influenced by many factors besides the level of the noise, such as general health status and attitude to the source of the noise. Historically, the most commonly-used method of assessing annoyance due to aircraft noise has been the use of questionnaire surveys with local residents. Early studies of that type investigated public perceptions of aircraft noise around airports; those studies revealed that aircraft noise causes relatively high levels of reported annoyance, and that increasing noise levels are correlated with increasing levels of reported annoyance, especially when the noise occurs at night (DfT 1992; Hume and Watson 2003; Morrell et al. 1997). In addition, the annoyance caused by aircraft noise was found to be greater in individuals who experienced sleep disturbance and disrupted communications, and who (a) defined themselves as noise sensitive; (b) expressed a concern about the health effects of noise; and (c) were afraid of aircraft crashes (Hume and Watson 2003; Morrell et al. 1997). However, one criticism of such studies is that, whilst 'annoyance' is readily reported for relatively loud noises (for instance, noises loud enough to cause awakening from sleep or to drown out the sound of a television), other effects may occur at lower sound levels and may not be perceived consciously as 'annoyance' – but may nevertheless contribute to degraded health or reduced wellbeing. By definition, of course, such low-grade annoyance may not always be appreciated or reported by the recipient, so is a more difficult phenomenon to research. Such considerations have led to the realisation that the effects of noise are not caused simply by the noise itself, and annoyance is not simply a function of the amplitude of a sound (DfT 2007; Fidell 1999; Hume and Watson 2003; Maris et al. 2007).

The fact that annoyance is affected by the 'social context of noise exposure' explains why conventional noise measures such as L*eq* have often failed to predict individual responses to noise; a vast range of other, psychological and sociological factors are involved whose influences are not captured by the use of a single measure (Maris et al. 2007). Fear of air accidents is one factor that may increase sensitivity to aircraft noise, leading to increased reporting of annoyance (Morrell et al. 1997; Reijneveld 1994). Negative affectivity (a bias towards focusing on negative features of the environment and to report negative experiences) may be another factor increasing the number of reports of annoyance. Annoyance due to aircraft noise may also be exacerbated by physical or mental illness, stress and anxiety – phenomena that may be especially distressing at night and thus preferentially reported. A survey of residents in the vicinity of London Heathrow Airport found that individuals reporting experiences of sleep disturbance, depression, irritability and tinnitus tended to be those who also reported high levels of annoyance due to aircraft noise (Tarnopolsky et al. 1980). Even a person's expectations about future noise levels may contribute to the annoyance they experience as a result of current aircraft noise events (Hatfield et al. 2001). Thus a study of residents in the vicinity of Amsterdam Schiphol Airport revealed the increased prevalence of severe annoyance due to aircraft noise compared with previous studies at the same airport and elsewhere: a result that was interpreted as an increase in local sensitivity to noise – together with greater concerns about safety – accompanying the contemporaneous debate about the expansion of the airport (Hume and Watson 2003). The task of understanding the effects of aircraft noise in terms of levels of annoyance is further complicated by the fact that individuals affected by noise may take measures to mitigate those effects: by installing noise-attenuation materials (such as sound insulation and double- or triple-glazing) in their homes; by complaining to airport operators and environmental health authorities; by instigating civil litigation procedures; by canvassing and obtaining media support; by organising political opposition, including protests; or by relocating (Gillen and Levesque 1994; Hume and Watson 2003).

Whilst noise-averse behaviours may reduce an individual's noise exposure, thereby representing factors that must be taken into account when attempting to evaluate levels of annoyance in the vicinity of airports, they may also serve as proxies reflecting the reduced wellbeing and quality of life that may result from exposure to aircraft noise. Consequently, researchers have investigated the extent to which such noise-averse behaviours are correlated with aircraft noise. In particular, the complaints made by individuals to airport operators, to regulators or to other authorities may offer valuable insights into the annoyance experienced (or at least reported) by people in the vicinity of airports (Hume et al. 2003). Noise-related complaints represent a common response to aircraft noise nuisance; furthermore, data about complaints are often recorded in a standardised form and may be referenced – using a Geographic Information System (GIS) – to particular locations. Thus modern noise monitoring systems coupled with information about aircraft movements and about the distribution of noise-related complaints offer researchers valuable opportunities to compare actual noise levels (and their timings) with the overall complaint-responses of local communities. Another reason why it is valuable to understand the distribution and behaviour of individuals who complain about aircraft noise is that restrictions on airport operations and infrastructure development are largely driven by complaints and by other forms of opposition from the communities affected. Hence a strong incentive exists to investigate 'complaint behaviour' and the links between annoyance, complaints and reduced health and wellbeing (Hume et al. 2003). One area of debate is whether individuals who complain about aircraft noise represent the population at large, or whether disproportionately vocal groups dominate debates about aircraft noise nuisance. This is a complex question: those who complain about aircraft noise constitute a self-selected group that undoubtedly includes people whose lives are genuinely blighted by aircraft noise as well as 'serial complainers' who may readily complain about aircraft noise regardless of the actual level of nuisance they suffer. To determine the extent to which noise complaints reflect the views of the wider population, data about complaints must be compared with broader social survey data. Hume and Watson (2003) have argued that, on balance, the reporting of aircraft noise nuisance tends to be dominated by serial complainers and pressure groups; those individuals and organisations may have disproportionate power in terms of their ability to influence decisions about proposed increases in airport traffic and planning applications for airport infrastructure development.

Given their political nature, how useful are noise-related complaints for understanding the effects of aircraft noise on the health and wellbeing of individuals in the vicinity of airports? Some authors have argued that, since complainers form only a subset of those actually affected by aircraft noise, they inevitably produce underestimates of the nuisance caused in a given community. Thus Borsky (1979) and Luz et al. (1983) have argued that aircraft noise complaints are not adequate measures of the community response to aircraft noise as they fail to represent the true extent of the problem; Hume et al. (2003) have also drawn attention to the fact that many individuals 'put up with' aircraft noise nuisance rather than make complaints. An alternative point of view is that complainers form a small group of people who are hypersensitive to aircraft noise nuisance (and thus not representative of the wider population), with the consequence that studies of noise-related complaints may actually overestimate the nuisance caused to the community as a whole. Whichever of those views is correct, many authors have argued that complaints can provide useful insights into the problem of aircraft noise nuisance (Bronzaft et al. 1998; Gillen and Levesque 1994; Hume et al. 2003; Stockbridge and Lee 1973). Hume and Watson (2003) have argued that a well-structured complaint system used in conjunction with an advanced noise monitoring system can provide useful insights into the relationship between aircraft noise and community annoyance – and the ways in which they might be more effectively managed. Currently, noise complaint systems vary widely in their quality and accessibility; consequently, many airport operators could potentially improve their data about noise nuisance – as well as their community relations – by providing an effective channel through which individuals are able to express their opinions about aircraft noise (see below).

Once high-quality information about noise-related complaints is available, some important trends may emerge from analysis of those data. Gillen and Levesque (1994) have demonstrated that, at Pearson International Airport, the number of complaints increased with the increased numbers of flights. At Manchester Airport, Hume et al. (2003, 61) found that:

- The number of noise complaints increased with the amplitude of the sound, with twice the number of complaints for overflights of 100dB(A) L*max* compared with those of 65dB(A) L*max* (measured outdoors).
- The number of complaints increased with increased media coverage of airport-related issues, including coverage of proposed airport infrastructure developments.

- The number of complaints varied diurnally, with twice the number of complaints about noise between 2300 and 0700 hours, with the most noise-sensitive time being between 0000 and 0100 hours, and with the least noise-sensitive time being between 1400 and 1500 hours.
- The number of complaints also varied on a weekly and monthly basis (after correcting for variations in traffic levels), and
- Most people who complained did so only a few times, whilst a few individuals (serial complainers) accounted for a large proportion of the complaints.

In a further study at Manchester Airport, Hume et al. (2003) found that night flights (2300–0600 hours) caused on average nearly five times as many complaints as flights during the rest of the day. In that study, those authors found that the time at which aircraft noise was most likely to result in noise complaints was between 0100 and 0200 hours, and the lowest between 0800 and 0900 hours (Hume et al. 2003). Such insights have important implications for the management of aircraft noise nuisance by airport operators and policymakers.

Sleep Disturbance

One of the most frequently-reported effects of aircraft noise is interference with sleep patterns (DfT 1992; 2003b; Hume and Watson 2003). Sleep disturbance is one of the most common causes of complaint from individuals who are subjected to aircraft noise at night; it often results from the noise of aircraft engines and airframes during approach, landing, take-off and departure – and also during certain ground operations such as engine testing – during the hours of night. (It is important to bear in mind that this is a generalisation; some people sleep during the day.) There are several aspects to the problem of sleep disturbance: it includes the experiences of being unable to get to sleep, of being awakened from sleep (once or repeatedly), of having sleep of reduced quality, of having difficulty awakening in the morning (after insufficient sleep) and of feeling inadequately rested after sleep. Such experiences are not exclusive; an individual may suffer some or all of these forms of sleep disturbance as a result of aircraft noise. Nor are these experiences necessarily constant, night after night, because individuals who are exhausted after a series of sleepless nights may then sleep through considerably more noise during the subsequent night. Overall, however, sleep disturbance – either alone or in combination with other exacerbating factors – may cumulatively lead to stress, anxiety, anger, frustration and exhaustion. The fact that adequate sleep (in terms of its duration, continuity and depth) may be regarded as a fundamental human right – and thus a prerequisite for human health and wellbeing – has been established by a ruling of the European Court of Human Rights following a legal challenge to the UK Government by noise-affected residents in the vicinity of London Heathrow Airport. Sleep disturbance, therefore, can be a profound source of misery for those whom it affects.

Many studies have investigated the relationship between aircraft noise and sleep disturbance; those studies have demonstrated that noise reduces the overall duration of sleep and alters the pattern of sleep stages during the course of the night (DfT 2003b; Hume and Watson 2003). Noise causes brief arousal of the central nervous system during all of the stages of sleep, as measured by electroencephalograph (EEG) recordings of brain activity (Carter et al. 1994a; 1994b; Whitehead et al. 1998). Studies by Wilkinson (1984) and Griefahn and Muzet (1978) demonstrated that noise at night may reduce both deep (slow-wave) sleep and the total duration of sleep. In addition, Morrell et al. (1997) showed that high levels of aircraft noise are associated with various sleep problems: delayed sleep onset, increased awakenings from sleep, sleep loss, premature awakening at the end of sleep and reduced quality of sleep. In addition to the effects of aircraft noise during periods of (attempted) sleep, further effects may occur the following day when individuals may perform sub-optimally both mentally and physically. Furthermore, some authors have argued that sleep disturbance due to noise – especially if chronic – could potentially cause or exacerbate serious illnesses, including some cardiovascular, endocrine and immunological diseases, although many uncertainties remain about the causal mechanisms of such effects (Carter 1996; Health Canada 2001; Health Council of the Netherlands 1999; Hume and Watson 2003). One of the reasons for the difficulty in accurately evaluating the relationship between aircraft noise and sleep disturbance is that, historically, many studies of the effects of noise-induced sleep disturbance have been undertaken in sleep laboratories rather than during field studies in subjects' homes. Once field study data became available, some discrepancies were observed between laboratory and field studies, with greater levels of sleep disturbance being reported in the laboratory studies (Hume and Watson 2003). Such a finding had potentially important implications for the management of aircraft noise because it indicated that subjects were more tolerant of noise than had previously been thought.

To clarify this issue, and to inform policymaking about night-flying operations, a major field study was undertaken in the UK during the early 1990s (DfT 1992). That study indicated that aircraft noise levels below 80dB(A) L*max* outdoors (which is equivalent to 55dB(A) L*max* indoors) caused little sleep disturbance, whilst higher noise levels produced approximately a 1 in 75 chance of being awakened from sleep (DfT 2003b, 31; Horne et al. 1994; Hume and Watson 2003). Perhaps surprisingly, that study found that aircraft noise was a relatively minor cause of sleep disturbance, with about 5 per cent of reported awakenings being attributed to aircraft noise, whereas domestic and other non-aircraft factors more frequently caused sleep disturbance. However, the study also revealed consistent, subtle effects of aircraft noise on the stages of sleep, although the significance of those effects was unknown (DfT 2003b; Hume et al. 1998). Similar results emerged from research with subjects in the US; except in the noisiest locations, awakening in the home due to aircraft noise was shown to be a relatively rare event and was only weakly correlated with the actual noise level (Hume and Watson 2003). Some caveats should be included here, however. As with any study that uses averaging techniques, the findings presented above reflect mean values and may obscure considerable individual variations in noise sensitivity and sleep disturbance. In fact, in the major UK study commissioned by the DfT (1992), large variations (of around 2.5 times) in the degree of sleep disturbance were found between individual subjects. Therefore, caution is required when interpreting values such as the threshold for noise-induced awakening, which is often quoted as being a level of 55-60dB(A), because changes in the stage of sleep – and minor arousals without perceived awakening – may occur at much lower noise levels (Hume et al. 2003). Furthermore, intermittent noise may cause greater sleep disturbance than continuous noise, including a reduction in slow-wave sleep, and it may result in poor performance and low mood the following day (Carter 1996; Ohrstrom and Rylander 1982). Fidell et al. (1995) also reported that individual noise intrusions are more closely related to annoyance and awakenings than are long-term, ongoing noise exposure – although, again, in reality, the situation is more complicated with additional confounding factors such as the familiarity of the sound (DfT 2003b).

More recently, the UK Government has reviewed the potential adverse effects of night-time aircraft noise (Porter et al. 2000). Despite concluding that no firm scientific evidence of clinically significant health impairment was found as a result of night-time aircraft noise, the UK Government could not deny the possible existence of cause-effect relationships (see also Health Canada 2001; Health Council of the Netherlands 1999). Various effects of night-time aircraft noise were identified in that review, including: (a) immediate physiological responses due to noise events that could lead to acute annoyance; (b) total night effects as a sum of the immediate responses, such as sleep reduction and fragmentation; (c) next-day effects, including increased sleepiness and reduced performance, causing perceived sleep disturbance, increased tiredness and annoyance; and (d) chronic effects that may constitute a deterioration of physical and mental health, with accompanying chronic annoyance and reduced quality of life (Hume and Watson 2003; Porter et al. 2000). Many confounding variables were also identified, such as attitudes to the noise source – which could strongly influence the relationship between night-time aircraft noise and effects on health and wellbeing. Yet despite the difficulties involved in demonstrating cause-effect relationships between night-time aircraft noise and the various dimensions of sleep disturbance, few people would deny that adequate sleep is an important aspect of human wellbeing. Whilst evidence from field studies suggests that aircraft noise results in only limited sleep disturbance for most individuals, and that – in most situations – it would be difficult to deprive people of sleep to the extent that unequivocal health effects became apparent, the notion that adequate sleep is important for wellbeing remains compelling.

Other Effects

Aircraft noise may have other, detrimental consequences for wellbeing. In particular, aircraft noise, especially if particularly loud or unexpected, may cause stress and anxiety – either directly or due to sleep disturbance. In turn, stress may be associated with reactions of alarm, resistance and exhaustion; it may also be experienced at varying levels ranging from low to severe (Hume and Watson 2003; Oken 2000; Selye 1956). Stress and anxiety may have important implications for health since they are risk factors for infection, cardiovascular disease (including hypertension), endocrine disease, gastrointestinal illnesses and immunological problems; in addition, some pre-existing illnesses may be exacerbated by fear and anxiety (Babisch 1998; 2000; Haralabidis et al. 2008; Health Canada 2001; Health Council of the Netherlands 1999; Ising et al. 1999; Morrell et al. 1997; Spreng 2000a; 2000b). The stress and anxiety caused by aircraft noise may be particularly acute for individuals who are more sensitive to noise or for those who are especially vulnerable to harm or illness. Exposure to noise can cause the release of 'stress hormones' (such as cortisol); Maschke et al. (1993) reported that subjects exposed to aircraft noise of 55-65dB(A) L*max* were found to have raised blood cortisol levels. Unsurprisingly, given the complexity of the physiological response to noise and the existence of a vast range of confounding psychological and sociological variables, many uncertainties remain about the mechanisms by which aircraft noise may generate stress, anxiety and disease (Babisch 1998; 2000; Health Canada 2001; Health Council of the Netherlands 1999; Morrell et al. 1997). A review of the impact of environmental noise on mental health by Stansfeld et al. (2005) found that exposure to high levels of noise may be associated with mental health symptoms and, possibly, with raised anxiety levels and increased consumption of sedative medication. Stansfeld et al. (2005) also acknowledged that persistent or repeated exposure to aircraft noise may reduce quality of life or wellbeing without necessarily inducing depression or anxiety; therefore, those authors called for more comprehensive studies of environmental noise, including the wider use of standardised methods to measure mental health outcomes and related physiological outcomes, such as hormone levels.

Recent research has indicated that chronic exposure to aircraft noise may impair children's learning, including their reading comprehension (Stansfeld et al. 2005). In the Road Traffic and Aircraft Noise Exposure and Children's Cognition and Health (RANCH) study, researchers examined the exposure-effect relationships in children aged nine and ten years; specifically, the relationship between chronic exposure to noise and impaired cognitive function, health and annoyance was examined. The findings of the RANCH study suggested that aircraft noise exposure was related to impaired performance in reading comprehension and recognition memory. In particular, the reading age of children exposed to high levels of aircraft noise was found to be delayed by up to two months in the UK for a 5dB(A) change in noise exposure. The results of that study indicated that long-term aircraft noise exposure impairs the development of children's reading skills, and that schools exposed to high levels of aircraft noise are not healthy educational environments. Similar effects of aircraft noise on the development of children's reading ability were found across three European countries. Long-term aircraft noise exposure was also found to increase the level of annoyance in children; such a stress response implies reduced wellbeing and a lower quality of life. The authors concluded that aircraft noise is an environmental stressor and that exposure to aircraft noise can impair children's health and their cognitive development (Stansfeld et al. 2005). To summarise the other effects of aircraft noise on wellbeing and quality of life, therefore, research to date has indicated that aircraft noise may exacerbate – and possibly cause – cardiovascular and mental illness, and may hinder children's cognitive development, and that further research is required to investigate those issues in greater detail.

Thus aircraft noise has a variety of complex, interrelated impacts on human health and wellbeing. Yet, despite the considerable efforts of researchers, many studies in this area remain contradictory, ambiguous or inconclusive. Porter et al. (1998) evaluated the degree to which knowledge about the potential health effects of environmental noise could influence the noise targets and standards set by the UK Government. However, despite finding evidence for several potential effects of environmental noise on health, mainly in relation to annoyance and sleep disturbance, Porter et al. (1998) concluded that, due to the confused state of knowledge about those effects, it was not possible to derive evidence-based health standards. As Hume and Watson (2003, 66) stated:

> It seems that we have not moved far from the conclusions reached by an exhaustive review by Miller in 1974, in which the only conclusively established effect of noise on health was that of noise-induced hearing loss, though it was accepted that noise can disturb sleep, be a source of annoyance, interfere with communication and performance at complicated tasks, adversely influence mood and disturb relaxation. In short, the author concludes with the lines 'noise can affect the essential nature of human life – its quality'.

Yet, despite the apparent lack of consensus amongst researchers about the precise impact of aircraft noise, the issue remains highly significant in relation to the operation and development of airports. The location of airports in, or near, urban areas almost inevitably involves inflicting a noise burden on the local community – and often on marginalised or vulnerable groups who are least able to mitigate the effects of that noise (Hume and Watson 2003; Maris et al. 2007). In addition, for a variety of reasons, people who live in the vicinity of airports are becoming increasingly intolerant of aircraft noise and its effects. Consequently, the management of aircraft noise is an important and growing challenge for airport operators, airlines, air traffic management (ATM) service providers, aircraft and engine manufacturers, regulators and planning authorities. In the following section, various ways in which the impact of aircraft noise can be reduced are discussed.

Reducing the Impact of Aircraft Noise

The growth in demand for air transport is projected to outstrip improvements in aircraft noise reduction technology until at least 2030 (Thomas and Lever 2003; see Chapter 2). Whilst the noise associated with individual aircraft movements is tending to decline due to technological and operational improvements, the frequency with which people are being over-flown by aircraft is increasing (DfT 2007). In addition, new patterns of demand for air transport – such as those resulting from the expansion of 'just-in-time' delivery strategies – mean that commercial pressures to operate more frequent flights, including flights during the hours of night, are intensifying. Another factor in the changing distribution of noise nuisance is the new patterns of operation that have accompanied the growth of 'low-cost' carriers (LCCs), which frequently fly from secondary airfields and which may thereby increase the noise levels experienced in places where traffic levels were previously much lower. One outcome of such trends is that, in the absence of additional environmental (or other) restrictions, the incidence of noise nuisance around many airports is expected to increase. Furthermore, in addition to the projected increase in frequency of flights, public tolerance of noise nuisance is generally decreasing as levels of affluence increase in many parts of the world, because rising affluence is accompanied by rising expectations in relation to quality of life, health and wellbeing. For reasons such as these, Thomas and Lever (2003, 110) have argued that the nuisance caused by aircraft noise 'is, and will remain, the single most significant local environmental impact resulting from the operation of airports'. Nor is the issue of aircraft noise confined to areas in the vicinity of airports and their associated flight-paths; increasingly, individuals and organisations complain about noise from aircraft at much greater distances (and altitudes) from airports, including nuisance in particularly noise-sensitive places (such as National Parks) where *any* noise generated by aircraft may be perceived as an intrusion into an otherwise tranquil environment (Miller 2007). Therefore, a significant challenge for the air transport industry is to meet the increasing projected demand for air travel whilst at the same time limiting or reducing the number of people exposed to the impacts of aircraft noise (Smith 1992; Thomas and Lever 2003). In this section, various ways in which the impacts of aircraft noise may be managed are discussed.

Worldwide, many different aviation noise policies and noise-abatement measures are in use (Girvin 2008). However, a starting point in the management of the impact of aircraft noise is the production of an accurate, comprehensive assessment of the nature, severity and distribution of that impact. Such an assessment indicates the nature of the noise burden to be mitigated, which is likely to be substantially different for different airports. As the preceding discussion has suggested, the problem of aircraft noise nuisance consists of two main elements: (a) the size and distribution of the population exposed to aircraft noise, which is typically defined as the population residing within a given noise contour; and (b) the degree of nuisance caused to those individuals as a result of their experiences of aircraft noise. It is worth emphasising that these two elements should be understood – and perhaps even managed – separately, because they involve different issues and their mitigation requires qualitatively different approaches. (However, this is not to deny the importance of adopting an integrated approach to the management of aircraft noise impacts – and indeed to the management of aviation environmental impacts more generally.) Yet the discussion presented above has emphasised the difficulties in characterising both of these elements; the first (noise exposure) is problematic because it is highly dependent upon the noise measure selected, whilst the second (the nuisance caused to individuals) is almost impossible to quantify since it is highly subjective and is influenced by a wide range of other factors. Nevertheless, the management of aircraft noise impacts should ideally be based on the most rigorous, impartial knowledge available, which generally means independently-derived knowledge obtained using standardised, verifiable scientific methods – even if those methods are sometimes relatively crude. Thus standardised noise monitoring and modelling techniques should be used to the extent possible in assessing aircraft noise in the vicinity of airports (see above). Of the two main elements that comprise the problem of aircraft noise (noise exposure and the experience of nuisance), noise monitoring and modelling relate primarily to the former, although a clear picture of aircraft noise exposure is a useful and important basis for understanding the latter.

Noise monitoring and modelling may be undertaken for various reasons: (a) to characterise the occurrence of aircraft noise at particular locations; (b) to inform and verify noise modelling exercises; (c) to investigate the degree to which aircraft operators (pilots and airlines) adhere to noise-related restrictions or agreements as they fly particular routes; (d) to provide data as a basis for the effective management of noise impacts; and (e) to inform dialogue between airport operators and neighbouring communities, since the implementation of an effective noise monitoring programme may be regarded as a crucial, initial step in the process of demonstrating that an airport operator is committed to the reduction of aircraft noise impacts. Since the problem of aircraft noise nuisance varies between airports, the type of data collected and the way in which those data are analysed and reported may differ according to local needs and priorities; nevertheless, as far as possible, the methods used for collecting, analysing and reporting data should be standardised and transparent. Noise monitoring typically occurs at specified points at or around an airport; those points may represent particularly noise-sensitive locations or they may be selected to allow the noise associated with particular arrival or departure routes to be monitored. Noise monitoring typically involves the collection of several types of information at each point: the timing and duration of individual noise events; the peak sound level occurring during each overflight; the duration of sound levels above a given threshold; and possibly the frequency of the sound. Ideally, those data are then input to an advanced noise and track-keeping monitoring system in which they are combined with other information (including radar data) about each aircraft movement: flight number, aircraft type, runway, arrival or departure route, lateral and direct distances from the threshold, height, the origin or destination airport and any deviations from the planned track. In turn, that information may be cross-referenced with flight schedules and with authoritative databases of airframe and engine data, meaning that precise airframe-engine combinations can be specified for each noise event. In addition, information about the location of sources of complaints about particular noise events may be incorporated into the database.

Those accumulated data may then be used in several ways: (a) to identify particular times during the day and night when noise events are concentrated and to correlate those events with traffic levels or with particular ATM procedures; (b) to identify the quietest and noisiest aircraft, airframes, engines, airlines or even (in principle) flight crews; (c) to assess the accuracy of track-keeping by flight crews and any relationship between inadequate track-keeping, noise-related complaints and noise levels at the surface; (d) to compare the noise burden associated with different arrival or departure routes, or with operations using different runways; (e) to investigate a range of other effects, such as the possible effects of ambient weather conditions on the noise levels received at the surface (if suitable meteorological data are available); (f) to analyse the factors that may lead to noise-related complaints on one occasion but not on another; and (g) to investigate the effectiveness of noise abatement procedures. Furthermore, as well as the capacity to investigate individual aircraft movements, a noise monitoring system should provide summary data, such as information about average noise levels for particular routes or particular runways. Thus advanced noise monitoring systems can potentially yield a wealth of information about aircraft noise in the vicinity of an airport, and they may reveal important patterns in the spatial and temporal distribution of aircraft noise exposure. As the preceding discussion has emphasised, measurements of exposure to aircraft noise – even those made using modern noise monitoring techniques – capture only part of the problem of aircraft noise nuisance. Yet those measurements may be supplemented by a wealth of information obtained from other sources;

that additional information relates to the subjective experience of noise nuisance as reported by individuals who are affected by the noise. Airport operators collect information about their neighbouring communities – and about levels of concern associated with aircraft noise – in a variety of ways, including records of community complaints, social surveys, public consultation exercises and analysis of media reports. All of those sources may provide useful information about aircraft noise nuisance, although they require careful interpretation.

Records of community complaints may indicate occasions when individuals have been pushed beyond a threshold of tolerance by a noise event, although they may also reflect a multitude of other factors (Hume et al. 2003). Clusters of noise complaints suggest that an individual aircraft movement has been more widely perceived as being excessively noisy and they may alert an airport operator to an overflight that requires further investigation. It is worth noting that, whilst most noise-related complaints usually relate to loud overflights, they may also be prompted by other events: aircraft flying 'too low' or in unusual attitudes or configurations; aircraft apparently flying off-track with respect to published arrival or departure routes; aircraft initiating 'go-around' manoeuvres whilst on approach to land; wake turbulence following an overflight; and even the smell of unburned aircraft fuel. Local geographical factors may also influence the spatial and temporal distribution of noise-related complaints at a given airport, because operations from different runways may affect different-sized populations and different social groups – factors that in turn influence the number of individuals with a propensity to complain about aircraft noise. Studies of noise-related complaints reveal that some individuals (serial complainers) complain regularly about aircraft noise, perhaps because they experience particularly severe nuisance, because they are particularly sensitive to noise, or because they have a particular agenda with respect to the operation and development of the airport (previously discussed). Whilst the views of serial complainers are, in principle, as valid as those of any other stakeholder, such individuals may undermine their case through indiscriminate complaints about routine noise events over which an airport operator may have limited control. Even despite the activities of serial complainers, the analysis of noise-related complaints may still yield useful information to airport operators, especially where that analysis reveals spatial or temporal changes in patterns of complaints (especially an underlying trend). Whilst it might seem counterintuitive that an airport operator would encourage local residents to complain about aircraft noise, in fact such a strategy is good practice, both for the purpose of improving community relations and for acquiring information about local residents' perceptions of aircraft noise nuisance. An airport should therefore have a well-designed, clearly-publicised complaint procedure and the data gathered should be analysed systematically in order to understand the exact cause of complaint in each instance.

Social surveys represent another means of gathering information about aircraft noise nuisance. Carefully designed social surveys may provide detailed information about local residents (both complainers and non-complainers), including the variety of concerns held by individuals within particular social groups. Social surveys may be used to create baselines – by assessing the level of noise nuisance that might be tolerated by various social groups – and then to evaluate any changes in the perceptions held by individuals over time, or in response to particular events (such as airport infrastructure developments or the introduction of noise abatement procedures). Information gained from social surveys may be supplemented by other data gained during public consultation exercises, which typically form part of the airport environmental impact assessments (EIAs) accompanying major infrastructure development proposals; public consultation exercises may also accompany the implementation of major new or revised operational procedures such as the redesign of airspace. Public consultation exercises may be useful to the extent that the relevant stakeholders are properly involved, that the exercise is transparent and accountable, and that the process offers a realistic prospect of stakeholder concerns influencing decision-making. Ideally, a public consultation process will reveal the extent of community support for, or opposition to, a given proposal (such as the introduction of noise abatement procedures); the aspects of that proposal that require revision; the likely impacts of any changes in operations at a given airport; and any other aspects of airport operations that cause concern. One further way in which airport operators can obtain information about the impacts of aircraft noise nuisance is through the analysis of media coverage. Media articles relating to aircraft noise nuisance (or other aviation issues) may raise awareness of aircraft noise amongst local residents, revealing issues of concern in different areas and potentially identifying the reasons for increasing numbers of noise-related complaints. The approaches and techniques mentioned above provide ways in which airport operators may better understand the nature of the noise nuisance problem, in order to manage that problem more effectively. How, then, can airport operators and policymakers respond to the problem of aircraft noise nuisance? A variety of measures is available to reduce both the number of people exposed to aircraft noise and their experiences of noise nuisance (Thomas and Lever 2003). As with other aviation environmental issues (Chapters 3 and 4), approaches to the management of the impact of aircraft noise focus on various technological, operational and policy options. Below, those main types of option are discussed in turn.

Technological Options

Significant advances have been made in aircraft engine and airframe technologies over the last three decades; those advances have focused on reducing the 'noise at source' generated by individual aircraft. Notably, the development of high-bypass-ratio modern turbofan engines, improvements in airframe design and the use of advanced sound-insulating materials have drastically reduced the sound levels generated by commercial civil aircraft and, in turn, the size of the noise contours associated with aircraft movements. Unfortunately, however, the potential to achieve further improvement is declining as the industry has become technologically mature, and further, incremental improvements in aircraft noise performance are increasingly expensive to achieve (Collins et al. 2006; Thomas and Lever 2003). In particular, the impressive progress made in developing quieter engines means that reducing airframe noise is becoming a task of equal significance – although one that could be ultimately more challenging (Smith 1992). Of particular concern for the management of other environmental issues – especially in relation to climate change – is the fact that technological improvements in the noise performance of aircraft may increasingly be achieved at the expense of fuel efficiency (as acoustic linings installed in aircraft incur a weight penalty, thus compromising fuel economy). Therefore, a trade-off exists between achieving further improvements in aircraft noise performance and reducing most gaseous emissions such as carbon dioxide (CO_2), as discussed earlier (Chapters 3 and 4). Even as new technologies are developed, the long lead-in times involved and the long in-service times of existing equipment mean that new technologies are generally incorporated into the global aircraft fleet very slowly. A cheaper alternative to purchasing (or leasing) new, relatively quiet aircraft is to upgrade particular components of the aircraft, such as the engines, with quieter versions, or to retrofit aircraft with noise-reduction devices. Previous attempts have been made to 'hush-kit' aircraft with such devices, although the benefits of hush-kitting in terms of noise reduction have been less impressive than might have been achieved by replacing noisier aircraft with new ones (Thomas and Lever 2003).

Operational Options

The noise generated by individual aircraft movements depends, in part, on how each aircraft is operated at a given airport. Furthermore, the overall noise exposure around an airport is influenced by the operational procedures used at that airport and in the surrounding airspace; thus it is affected by airline operating procedures and by ATM systems and procedures (Clarke 2003). Many of those factors could be revised in order to reduce the noise associated with aircraft operations, and various revised operational practices have been developed by aircraft operators, ATM service providers and regulators either (a) to reduce the noise associated with individual aircraft movements, or (b) to alter the distribution of that noise. Revised operational practices are designed for use in specific stages of the aircraft landing and take-off (LTO) cycle and, if combined, they potentially offer an integrated approach to noise abatement for the entire LTO cycle. Aircraft approaching an airport for landing may be directed along a noise preferential route (NPR) which is designed to avoid exposing noise-sensitive locations to unnecessary overflights. Inbound aircraft may fly a continuous descent approach (CDA) or a 'low-power, low-drag' (LP/LD) approach, both of which may reduce airframe and engine noise since they are designed to give flight crews greater control over the energy of the aircraft throughout the entire approach profile – ideally avoiding the need for inefficient level segments and needless fuel burn. Depending on the operational circumstances, during CDA and LP/LD approaches, flight crews may be able to make minimal use of speedbrakes and flaps, which generate considerable noise along with substantial parasite drag. (However, those two procedures differ in some important respects, including the thrust settings that are used during the approach, and so they may also differ in the extent to which they are effective at reducing noise during some parts of the approach.)

During the landing roll, flight crews may be requested to avoid using reverse thrust and instead to use the brakes to decelerate the aircraft (where that can be done safely); by so doing, a significant reduction in aircraft noise may be achieved, although this procedure may compromise the turn-around time at the airport due to the need to allow for brake cooling. During taxiing, aircraft may be directed to use the taxiways that result in the least noise being generated near to residential areas. If aircraft are required to wait at taxiway holding points, ATM procedures could ensure that this occurs in locations where noise propagation has least impact upon local residents (Thomas and Lever 2003). On-stand, aircraft may be powered by fixed electrical ground power (FEGP) supplies rather than by relatively noisy auxiliary power units (APUs). Engine testing can be minimised and should, where possible, be avoided during the hours of night. For departing aircraft, revised take-off procedures may be used. Reduced thrust take-off – which involves flight crews using a de-rating procedure – means that aircraft are operated at a lower thrust setting during the take-off roll and the early climb, with higher power settings being reserved (if required) for subsequent stages when the aircraft has climbed above the atmospheric boundary layer. In addition, two noise abatement departure procedures (NADPs) have been defined by ICAO; those procedures involve selecting particular combinations of thrust and flap settings in order to minimise the noise received at the surface during departure (Connor 1996). Inevitably, the use of reduced thrust during take-off and climb compromises the rate of climb of an aircraft, with the result that noise is generated at low altitude for a longer period. Therefore, a trade-off exists between the noise benefit of reduced thrust operations and that of a more rapid climb out of the LTO cycle. Another trade-off exists with the reduction of CO_2 (and most other) emissions in cases where the use of revised operational procedures, such as NPRs, requires aircraft to fly longer routes in order to avoid noise-sensitive areas – and thus increases fuel consumption and most emissions.

Policy Options

As is the case with the management of other aviation environmental issues (see Chapters 3 and 4), policy options may be categorised broadly as regulatory, market-based and voluntary approaches. The regulation of aircraft noise occurs globally through ICAO; the approach adopted by ICAO focuses on promoting incremental reductions in noise at source through the implementation of progressively stricter noise certification standards that must be met before new aircraft are permitted to enter into service, together with the introduction of operational restrictions of the noisiest aircraft (Franken 2001; ICAO 2006a). The fundamental premise of the ICAO approach is that regulation should seek to drive the development and adoption of quieter technologies and the gradual phase-out of noisier aircraft. In that way, the effects of the future growth in air traffic – and the increasing intolerance of aircraft noise nuisance within many communities – may be offset by continuous improvements in aircraft noise technology and by a rolling airline fleet replacement programme. Thomas and Lever (2003) have acknowledged that the regulation of aircraft noise by ICAO has been a major driver of the introduction into service of improved aircraft noise technology: (a) by setting noise performance targets that must be met before aircraft are licensed to fly, and (b) by introducing targets for the phase-out of older, noisier aircraft from the fleet. The regulation of aircraft noise by ICAO evolved in the context of increasing public opposition to noise nuisance in some parts of the world (see Chapter 1). The ICAO noise regulations (known as 'Chapters') are specified in Volume I of Annex 16 to the Convention on International Civil Aviation (the Chicago Convention of 1944). Various levels of stringency exist in those standards. The noisiest aircraft were designated as 'uncertificated'; operation of those aircraft was prohibited in Europe from 1985. The Stage 2 regulations (which applied to so-called 'Chapter 2' aircraft) set out the standards applicable to jet aircraft designed before October 1977, whilst Stage 3 (which refers to quieter, 'Chapter 3' aircraft) defined the more stringent standards applicable to jet aircraft designed after that date. In September 2001, ICAO approved a further increase in the stringency of the standards, and new ('Chapter 4') requirements were introduced for new aircraft certificated from 1 January 2006 onwards, although the new standards represented only a modest increase in stringency – of $10dB(A)$ – over the Chapter 3 standards (Garvey 2001). Concurrently, the introduction of operational restrictions required that all Chapter 2 aircraft were retired from service by 31 March 2002 unless they were re-certificated to meet Chapter 3 standards. The debate about ICAO operational restrictions now focuses on the question of the timing of the phase-out of Chapter 3 aircraft from service (unless they are re-certificated to meet Chapter 4 standards; Depitre 2001). However, no phase-out programme for Chapter 3 aircraft has yet been introduced.

Since, at the global scale, nations differ in their capacity to comply with the ICAO regulations – and because air transport represents an important driver of economic and social development in many parts of the world – ICAO regulation of aircraft noise is a sensitive issue as it could potentially compromise sustainable development (Garvey 2001; see Chapter 6). Thus, in industrialised countries, new civil aircraft now easily exceed the ICAO Chapter 4 standards whilst older variants operating in developing countries may not comply with those standards. In an attempt to ensure that operating restrictions do not impact disproportionately upon developing countries, ICAO has called instead (in its Resolution A33-7) for the adoption of a 'balanced approach' to the management of noise around airports, involving four principal elements: (a) the reduction of noise at source; (b) land-use planning; (c) noise-abatement operational procedures; and (d) operational restrictions (EC 2002; ICAO 2006a; Thomas and Lever 2003, 103–104, 106). ICAO has urged UN Member States not to introduce any operating restrictions of Chapter 3 aircraft before making full assessments of all available options for noise mitigation at any given airport in accordance with the balanced approach (Thomas and Lever 2003). The ICAO Assembly has also specified various safeguards to be applied if restrictions of Chapter 3 aircraft are introduced: for example, restrictions should specifically address the problem of noise nuisance at the airport concerned, and the special circumstances of aircraft operators in developing countries should be taken into account.

For some campaigners, the apparently slow pace of progress made by ICAO in increasing the stringency of noise certification standards and in phasing-out older, noisier aircraft is a source of frustration. However, ICAO is accountable to all UN Member States and faces the task of balancing a range of significant – and perhaps conflicting – demands: to safeguard economic growth, to promote economic development, to maintain and develop critical air routes, to prioritise air safety and to ensure environmental protection. A regional-scale example of the problems involved in management such conflicting demands emerged in Europe with the accession to the European Union (EU) of countries in Central and Eastern Europe; those countries faced demands to expand their air route networks whilst simultaneously modernising their aircraft fleets in order to comply with noise standards at EU airports. The high costs of fleet renewal with modern aircraft meant that fleet sizes were reduced, with the consequence that national carriers had insufficient capacity to expand their route networks at a critical time when new markets were being opened (Thomas and Lever 2003). The complexity of international negotiations makes the development of the global regulatory framework for aircraft noise management a slow process; considerable time is required to undertake public consultations and to secure agreements between governments, non-governmental organisations (NGOs) and the air transport industry. The task of securing such agreements is complicated by the fact that, in many parts of the world, the management of aircraft noise nuisance is not a priority, and the capacity to modernise airline fleets is severely limited, whereas the demand for cheap, accessible air transport services is buoyant. In contrast, in other regions, such as Europe, both the capacity to undertake fleet modernisation and the levels of public opposition to aircraft noise nuisance are relatively high, creating strong incentives to address the issue. Yet, even in developing nations, there is still an imperative for fleet renewal – not least so as to allow airlines to access airports in developed countries. ICAO faces a significant challenge in attempting to resolve such tensions. On the one hand, attempts by ICAO to achieve acceptable compromises in negotiating noise regulations may mean long delays in implementing new certification standards and operational restrictions; furthermore, given the rapid growth of demand for air transport, such delays may lessen the effectiveness of new regulations. On the other hand, pressing ahead with regulations that fail to meet diverse national requirements could potentially lead to the adoption of weak standards that fail to provide people with adequate protection against the effects of aircraft noise – or alternatively with stronger standards that may be impossible for many airport operators to meet (Garvey 2001, 20; ICAO 2006a; Thomas and Lever 2003). Overall, the global regulation of aircraft noise requires ongoing processes of review, improvement and the spread of good practice in accordance with the principles of sustainable development (Garvey 2001; see Chapter 6).

Whilst the regulation of aircraft noise by ICAO occurs at the global scale – and focuses on the adoption of a 'balanced approach' to the management of aircraft noise nuisance – another regulatory framework has been developed within the EU (EC 2002; ICAO 2006a). Levels of public sensitivity to aircraft noise are especially high within the EU due to high population densities, high frequencies and widespread distribution of air services and very high expectations in relation to quality of life and wellbeing; people living in the vicinity of major European airports are some of the most noise-sensitive individuals in the world (EC 2002; Skogö 2001; Thomas and Lever 2003). Given those particular circumstances at the regional level, the EU has acknowledged that aircraft noise nuisance represents a critical obstacle to the development of the European air transport system. Following the refusal of the ICAO Assembly in 2001 to increase the noise stringency standards for Chapter 4 aircraft or to introduce a phase-out programme for Chapter 3 aircraft, the EU introduced a Directive (2002/30/EC) on the establishment of rules and procedures with regard to the introduction of noise-related operating restrictions at European Community airports (EC 2002). That Directive acknowledged that particular noise problems occur in the vicinity of some airports (termed 'city airports') that are located within large conurbations, and that those problems should be alleviated by allowing the introduction of more stringent rules at those airports (the city airports were defined as Berlin-Tempelhof, Stockholm Bromma, London City and Belfast City Airports; EC 2002). The Directive was intended to: (a) enable noise-sensitive airports to apply to take action to exclude the noisiest Chapter 3 aircraft; (b) establish a framework within which European airports would implement common noise-related charging regimes designed to encourage the use of quieter aircraft and discourage operations by noisier types; (c) establish a European framework within which operational restrictions would be applied to different aircraft; and (d) select appropriate mitigation measures, with the goal of achieving the maximum environmental benefit most cost-effectively (EC 2002; Thomas and Lever 2003). Additional EU legislation has been introduced in the Directive relating to the assessment and management of environmental noise (Directive 2002/49/EC; see above) which includes noise from air traffic in the vicinity of airports and which introduced new requirements for noise action plans to be produced and for noise performance targets to be set. In those Directives, then, the EU has established a regional regulatory framework that is designed to facilitate aviation growth, to take into account the differing needs of countries within the EU and to reduce the total number of people exposed to noise (EC 2002; Thomas and Lever 2003).

In addition to those frameworks at the global and regional scales, the regulation of aircraft noise also occurs at the level of individual airports. Noise-related operating restrictions have been introduced at airports worldwide. Most of those restrictions apply to night flying: airports may be required to cease operating at night (curfew) or may have limits imposed on the numbers and/or types of aircraft that are permitted to take-off or land during the hours of night (although some variation in the definition of 'night' occurs between countries and between airports). At some airports, restrictions of the use of particular runways are applied at noise-sensitive times and night flying restrictions may specify that operations may not exceed specified night-time noise contours. In some cases, the operators of major airports have agreed to, or have been forced to adopt, day-time movement capacity limits – often as part of a planning agreement to allow future infrastructure development. Another common type of constraint is a limit to the noise that can be generated over a specified period of time; once such a limit is reached, further growth may only occur through the introduction of quieter aircraft (Thomas and Lever 2003). Such airport-specific operating restrictions may be combined in various ways to suit local needs. For instance, at London Heathrow, Gatwick and Stansted Airports, a Night Restrictions Regime was introduced that limited the operations of the noisiest aircraft (as defined by their Certificated Noise Level). During the 'night quota period' (defined as 2330–0600 hours), all aircraft movements are restricted according to their Certificated Noise Level, based on a Quota Count (QC) system; in addition, during a slightly longer 'night period' (defined as 2300–0700 hours), the noisiest types of aircraft may not be scheduled to take-off or land (DfT 2003b).

In addition to the direct approaches described above, the regulation of aircraft noise occurs indirectly through land-use planning at and around airport sites (Koppert 1996). The use of land-use planning to mitigate aircraft noise nuisance may focus on the airport site itself, on adjacent developments, or on both, either through decisions about planning permission or through the use of zoning. Effective land-use planning at and around airports can potentially ensure that the minimum number of people live or work in areas exposed to high levels of noise; however, where land-use planning controls are inadequate – as in many parts of the world – residential properties may still be constructed close to airports, not least because the growth of airports may inflate local land values and thereby stimulate urban development. Furthermore, there is often the historical problem that many airports have long been located adjacent to pre-existing urban areas, with noise-sensitive locations in the vicinity of their runways and beneath their arrival and departure routes, or they may have the legacy of local developments that were approved prior to a period of recent, rapid airport expansion. Thomas and Lever (2003) have acknowledged that a tension exists between the desire of land-use planners to create a buffer zone around airports within which no population is affected by noise nuisance and the increase in land values that occurs due to easy access to an airport and which in turn stimulates development proposals. In extreme cases, where aircraft noise results in unacceptable impacts on an urban population and those impacts cannot be mitigated at the existing site, entirely new airports (such as Hong Kong International Airport) have been constructed in more distant or offshore locations; such new airports typically rely upon rapid surface transportation to provide access to the urban centre. Where the relocation of an airport is not feasible or desirable, aircraft noise may be mitigated indirectly through careful airport design, including the strategic construction and siting of airport buildings (as at Copenhagen Airport, where the domestic terminal building acts as a noise shield) and of engine-test facilities.

Another means by which aircraft noise nuisance may be mitigated is by compulsory buy-out and sound insulation schemes. Compulsory buy-out schemes apply in cases where residential properties are located within noise contours representing the limit of 'unacceptable' noise levels; however, they are now used relatively rarely since few properties exist in such close proximity to airports, and those properties tend to be owned already by airport operators. In contrast, the sound-proofing of homes, public buildings and business premises in close proximity to an airport is widely undertaken by airports as a way of mitigating the impacts of aircraft noise. As Thomas and Lever (2003) have acknowledged, sound-proofing may provide considerable relief of aircraft noise nuisance. However, its effectiveness is limited for several reasons: (a) sound insulation schemes require considerable expenditure; and (b) whilst sound levels are significantly reduced within insulated buildings with the windows closed, the benefits may be lost during the summer months when individuals tend to open windows and to spend more time outdoors. Nevertheless, the provision of sound insulation may be mandatory for airport operators, for properties located within specified noise contours.

One further approach to the management of the impact of aircraft noise involves the use of market-based measures, including noise-related charges or penalties such as the noise levy charge applied at Sydney Kingsford Smith Airport (Nero and Black 2000). That approach to noise reduction relies on the use of differential charges based on monitored noise performance and it aims to encourage the use of quieter aircraft and to discourage noisier operations. Noise-related charges or penalties may be adjusted to reflect varying sensitivity at different times of the day and night, thereby targeting the operations that cause the greatest nuisance to local residents – and also providing a degree of adaptability to local circumstances. If noise-related charges and penalties are imposed on the operators of aircraft that exceed agreed noise levels during take-off or landing, then an incentive may be created for flight crews to adopt the quietest possible operating procedures (Thomas and Lever 2003). In general, market-based approaches are based on the principle of providing actors (in this case, airports, airlines, ATM service providers and aircraft and engine manufacturers) with incentives to internalise the full social and environmental costs of their activities at the margin; thus they comply with the polluter pays principle (Ben-Yosef 2005; Nero and Black 2000). Other than noise charges and penalties, market-based measures have been little used to manage the impact of aircraft noise; consequently, they may have considerable potential to be used for that purpose, although there is currently no functioning market for noise (Ben-Yosef 2005; see also Brueckner and Girvin 2008).

The options and approaches described above represent practical ways in which aircraft noise nuisance may be reduced, limited or mitigated. Hence those options and approaches focus on the reduction of noise exposure: the number of people who experience aircraft noise and the levels of noise to which they are exposed. However, the discussion presented in this chapter has emphasised that, with respect to the impact of aircraft noise, noise exposure is only one consideration; the other major component of aircraft noise nuisance is the subjective experience of aircraft noise and the complex, variable responses of individuals to that experience. Therefore, efforts to manage aircraft noise nuisance should not be limited to measures to reduce noise exposure (Maris et al. 2007). Airport operators can voluntarily do a great deal to communicate more effectively with their stakeholders – including local residents who suffer noise nuisance – and to attempt to understand the true impacts of aircraft noise on local people. Through genuine community relations and stakeholder consultation exercises, and through investing time, effort and resources in communicating with local residents, airport operators may be successful in explaining their business to local residents in terms of economic, social and environmental costs and benefits. Furthermore, by genuinely listening and responding to the concerns of local residents and other stakeholders, airport operators could develop better relationships with their neighbouring communities and may be able to devise ways of sharing the benefits of their operations more equitably with those who suffer the impacts of airport operations. Airport operators could demonstrate leadership in that process by providing, for example, preferential employment opportunities for local people, or through sponsorship or education programmes that increase the employment prospects of people in low-income social groups in neighbouring communities (Thomas and Lever 2003). The broader interactions between aviation environmental issues – such as the impact of aircraft noise – and sustainable development are discussed in more detail in the following chapter.

Summary

The nuisance caused by aircraft noise is now, and is likely to remain, the single most significant local environmental impact resulting from the operation of airports (Thomas and Lever 2003). Whilst studies of the direct impacts of aircraft noise on human health remain suggestive but largely inconclusive, many aspects of human quality of life and wellbeing are clearly degraded by the annoyance and sleep disturbance that may accompany the experience of aircraft noise. This chapter has focused on the two main dimensions of aircraft noise nuisance: exposure to noise (which may be mitigated by means of a range of regulatory, market-based and voluntary measures) and the nuisance associated with the subjective experience of noise (which may at least be acknowledged and understood by airport operators as part of their broader community relations and stakeholder engagement processes). Whilst it may seem to be a trivial concern to those who do not suffer its effects, the issue of aircraft noise is highly important: aircraft noise affects the lives of millions of people worldwide, each day, with the consequence that aircraft noise nuisance is currently the main environmental constraint of growth at many major airports and one of the most critical issues influencing decisions about airport infrastructure development. Yet the problem of aircraft noise nuisance does not affect airports and their neighbouring communities in a uniform manner but instead varies depending on local circumstances. Some airport operators face severe noise-related constraints due to their location within – or in close proximity to – acutely noise-sensitive urban areas. For others, aircraft noise is hardly a concern, whereas the provision of reliable access to cheap air transport services is an overriding priority.

At the global scale, major challenges are faced by ICAO – and by other authorities and organisations – in managing and regulating aircraft noise. Marked spatial variations in sensitivity to aircraft noise, variations in the capacity of airlines to undertake fleet renewal, the relentless growth of demand for air transport and the technological maturity of the industry conspire to make the negotiation and agreement of noise-related regulations and operating restrictions extremely difficult. Some regions – notably the EU – have demanded greater flexibility to exert stricter control over aircraft noise; yet, in a global industry, the legislation enacted in one region affects all others. Concerns have been expressed that imposing stringent noise-related restrictions could disproportionately affect the airlines of developing countries, those with the least capacity to undertake fleet renewal. Therefore, in debates about the management of the impact of aircraft noise, the environmentalist and sustainable development agendas may collide. To some extent, this is true of all aviation environmental issues, but the conflict between the environmentalist and sustainable development agendas is revealed particularly starkly by the issue of aviation noise – because that issue may be regarded by some as one that primarily affects those affluent consumers in industrialised nations who most benefit from aviation-related growth and from the availability of affordable air transport. This chapter has attempted to show that the issue is more complex than such a view would suggest. Aircraft noise has intolerable effects on some people – especially on more vulnerable individuals – regardless of their country of residence, their affluence level or the extent to which they use air transport services. Consequently, the growth of the air transport industry will depend significantly upon achieving ongoing, substantial reductions in aircraft noise at source. Some of those reductions may be achieved through improvements in airframe and engine technologies, and aircraft noise may also be mitigated to some extent through the use of revised operational procedures. However, as is the case with other aviation environmental issues, technological and operational improvements are unlikely to offset the effects of the projected rapid growth of demand for air transport (see Chapters 3 and 4). Thus, with time, the capacity of many airports to meet the growing demand for air transport is likely to be severely constrained. In turn, such constraints could have profound implications in terms of lost opportunities for economic and social development. So a critical question remains: should aviation growth be constrained in order to keep noise – and other aviation environmental impacts – within acceptable limits, or could the intensifying environmental impacts associated with meeting growing demand for air transport be justified on other (economic and social) grounds? That question receives more scrutiny in the next chapter.

6   Air Transport and Sustainable Development

Introduction

Debates about environmental issues – particularly global environmental issues – are often framed within broader debates about sustainable development (Adams 2009). This has occurred with the ascendancy of the concept of sustainable development, which came to prominence as a result of several high-profile United Nations summits, including the United Nations Conference on Environment and Development (UNCED) held in 1992 in Rio de Janeiro and the World Summit on Sustainable Development (WSSD) held in 2002 in Johannesburg. Sustainable development has rapidly become the central organising principle in debates about how to manage environmental impacts whilst at the same time ensuring that the development of human economies and societies occurs in an equitable way (Adams 2009; Baker 2006; Dresner 2008; Elliott 2006). Many definitions of sustainable development have been proposed; most of them include the idea of balancing the economic, social and environmental costs and benefits of development, both for present and future generations. Thus proponents of sustainable development argue that economic development and environmental protection are not mutually-exclusive goals but should occur together in an integrated manner. Whilst such a view is attractive and seemingly benign, achieving such a reconciliation of economic, social and environmental concerns is far from easy and raises thorny problems of governance, power and justice. Ideas of sustainable development, then, are intrinsically political and cannot be divorced from considerations of participation, representation, transparency and accountability (Jacobs 1995; Redclift 1984). Notions of sustainable development also contain myriad assumptions – and often many contradictions – that ideally should be acknowledged and analysed critically. In relation to air transport and sustainable development, a central question is whether the growth of air transport should be constrained in order to keep aviation environmental impacts within acceptable limits, or whether that growth should be allowed and its intensifying environmental impacts justified on other (economic and social) grounds. That question is about values, attitudes and beliefs far more than it is about technologies and operating practices, important as those factors are. Moreover, it is a question that legitimately concerns many more people than those directly involved with the air transport industry.

Ultimately, however, the task of setting the direction of air transport (and broader) policy falls to decision-makers; policymakers must then attempt to reconcile that course of action with the principles and imperatives of sustainable development. Whilst the latter task represents a formidable challenge, it is nevertheless the only way in which a compromise between aviation growth (and economic growth more generally) and environmental protection may be found that is acceptable to the majority of people. Yet the term 'sustainable development' is notoriously slippery: 'it either means practically nothing to people, or practically everything' (Dresner 2008; Jacobs 1991; O'Riordan 1988; Porritt 2005, 22; SDC 2001b). Hence it is important to understand the idea of sustainable development and its implications for environmental management. In this chapter, the concept of sustainable development and its relevance to the management of aviation environmental impacts are explained. Ideas about sustainable development are complex and fiercely contested; below, the main principles of sustainable development are discussed, including the key principles of intergenerational and intragenerational equity. Crucially, poverty reduction is a central aspect of sustainable development – and this is one area in which air transport could potentially make a substantial contribution to the task of promoting sustainable development. Next, in this chapter, the relationship between air transport and sustainable development is examined in more detail by evaluating the rhetoric found in some high-level statements about aviation and sustainable development, and by considering the significance of the economic benefits of air transport. This is an important subject for those concerned with air transport management because aviation generates both strong economic growth and increasingly severe environmental impacts; the management of those environmental impacts – whilst essential – should not blight economic development where it is most needed. In this chapter, some of the ways in which air transport can be used to promote sustainable development, and some possible implications for policy and practice, are explored. In particular, poverty reduction emerges as an important aspect of the relationship between air transport and sustainable development, because the air transport industry has unique potential to contribute to poverty reduction worldwide. Overall, as the processes of globalisation intensify, and as concerns about global environmental change – especially climate change – become more acute, the imperative for air transport to operate on a basis that promotes sustainable development is becoming increasingly clear.

Sustainable Development

'Sustainable development' is a popular and commonly-used term, yet one that also arouses fierce debate and controversy because it is regarded as 'a meeting point for environmentalists and developers' (Dresner 2008, 70; Jacobs 1991; O'Riordan 1988). In an authoritative formulation by the World Commission on Environment and Development (WCED 1987, 43), sustainable development was defined as 'development which meets the needs of the present without compromising the ability of future generations to meet their own needs'. Although the term has since been fiercely contested and widely interpreted, most definitions of sustainable development retain the central idea of balancing the economic, social and environmental aspects of development, for present and future generations, and many also adopt a global perspective (Adams 2009; Baker 2006; Dresner 2008; Elliott 2006). The three core aspects of sustainable development – economic, social and environmental – are referred to as the three dimensions or pillars of sustainable development (Baker 2006; Ekins 2000; Figures 6.1 and 6.2). The same idea has also been expressed in terms of the 'triple bottom line': the notion that economic, social and environmental benefits and costs should be fully accounted for in evaluating human activities (Porritt 2005). Furthermore, as Baker (2006, 7–8, emphasis in original) has acknowledged:

> Sustainable development is a dynamic concept. It is not about society reaching an end state, nor is it about establishing static structures or about identifying fixed qualities of social, economic or political life. It is better to speak about *promoting*, not achieving, sustainable development. Promoting sustainable development is an on-going process, whose desirable characteristics change over time, across space and location and within different social, political, cultural and historical contexts.

The idea of promoting sustainable development implies that alternative futures lie before society; that those alternative futures may be envisioned; and that, through changes in attitudes and values, policy innovations, political transformations and economic restructuring, it is possible to work towards a future that is sustainable (Adams 2009; Baker 2006).

Figure 6.1   Sustainable development: linking economy, society and environment

Source: Adapted from Baker (2006, 8)

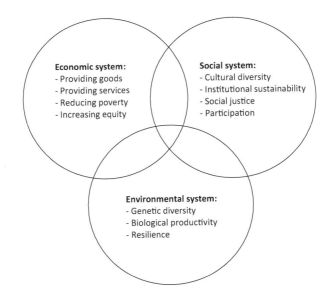

Figure 6.2   Some objectives of sustainable development

Source: Adapted from Elliott (2006, 13)

The idea of sustainable development came to prominence following the work of the WCED (1987) and, subsequently, as a result of several high-profile United Nations summits, including the United Nations Conference on Environment and Development (UNCED) held in 1992 in Rio de Janeiro and the World Summit on Sustainable Development (WSSD) held in 2002 in Johannesburg. Sustainable development became a popular concept because it seemed to offer a way to overcome certain critical issues facing policymakers. Previously, developers (who sought to promote economic growth and who wished to exploit natural resources to that end) and environmentalists (who sought environmental protection and who argued that economic growth was incompatible with that goal) had adopted conflicting, entrenched positions. Instead of focusing on the conflict between economic growth and environmental protection, the WCED argued that both of those goals are achievable. The WCED (1987, 8) stated:

> The concept of sustainable development does imply limits [to economic growth] – not absolute limits but limitations imposed by the present state of technology and social organization on environmental resources and by the ability of the biosphere to absorb the effects of human activities. But technology and social organization can be managed and improved to make way for a new era of economic growth.

Thus, by anticipating advances in technologies and changes in social organisation, the WCED (1987) was able to reconcile the requirement for environmental protection with the desire for economic growth. As a result, proponents of sustainable development have argued that economic growth and environmental protection are not mutually-exclusive but should occur together in an integrated manner. Since that time, sustainable development has rapidly become the major organising principle in debates about how to manage environmental impacts whilst at the same time ensuring that the development of human economies and societies occurs in an equitable way. The relevance of ideas about sustainable development has also been heightened by several other factors: (a) the realisation that conventional approaches to development have apparently failed, despite decades of development-related efforts and investment, given the persistence of poverty worldwide; (b) the emergence of postmodern critiques of conventional approaches to development and the subsequent impasse in development studies; (c) advances in the scientific understanding of the magnitude and pace of global environmental change, including biodiversity loss and climate change, and of the urgent need for more effective environmental protection; and (d) new ideas about governance, power, participation and the role of the state, especially in the context of rapid globalisation (Adams 2009; Schuurman 1993).

Yet, whilst the idea of sustainable development has become a central organising principle in debates about reconciling economic growth and environmental protection, it is important to acknowledge that the term is highly problematic and strongly contested (Adams 2009). Some commentators maintain that, despite the high-level visions articulated by the WCED (1987) and at the various UN summits, economic growth and environmental protection cannot be so easily reconciled (Adams 2009; Baker 2006; Dresner 2008; Elliott 2006; Jacobs 1991; O'Riordan 1988). In particular, concerns have been raised that the technological advances upon which the WCED (1987) vision depends are unlikely to offset fully the rate of economic growth. Hence technological advances that increase 'eco-efficiency' – which may be characterised as 'achieving more with less' – are an important part of promoting sustainable development but they are not sufficient to prevent devastating environmental degradation, given the magnitude and pace of economic growth and the severity of its environmental impacts (Adams 2009; Hajer 1996; Hartwick and Peet 2003; Pepper 1993; UK Government 2005). Moreover, the vision of sustainable development set out by the WCED (1987) involves profound changes in social organisation, yet some commentators have suggested that such changes are unrealistic and are unlikely to occur without unprecedented economic and political restructuring, if at all (Dresner 2008; Elliott 2006; Starke 1990). In particular, given recent trends of globalisation, the world economy is now dominated by multinational corporations with unprecedented power that are accountable to no single national government and that are able to exert strong downward pressure on both social and environmental standards (Adams 2009; Lash and Urry 1994). Some commentators have suggested that the discourse of sustainable development actually diverts attention from fundamental questions about the desirability of economic growth, about social development and about what constitutes wellbeing and quality of life (Eckersley 1992; Ponting 2007). Such concerns have led some authors to conclude that the term 'sustainable development' may be so vague as to be effectively meaningless: it is used by governments and businesses – including the most environmentally-destructive organisations – as a 'greenwash' to conceal a 'business-as-usual' approach to economic growth (Adams 2009; Jacobs 1991; Redclift 1987).

An indication of the complexity associated with ideas of sustainable development is given by the fact that 'strong' and 'weak' versions of sustainability have been articulated; those versions differ in the extent to which they permit natural capital (such as fossil fuels) to be drawn down and converted to human capital (such as education). Strong forms of sustainable development focus on environmental protection, which is regarded as a prerequisite for economic development, and they place strict limits on the amount of natural capital that may be converted to human capital. In contrast, weak forms of sustainable development prioritise economic development, which is viewed as a precondition for environmental protection, and they permit much greater – or even total – substitutability of human capital for natural capital (Adams 2009; Baker 2006). Other differences also exist between strong and weak versions of sustainable development. One important difference is that strong versions of sustainable development tend to seek economic and political restructuring such that wellbeing and quality of life are prioritised over economic growth *per se*, whereas weak versions tend to focus on increasing the absolute size of economies and they often include assumptions that additional, social benefits will accompany that growth. Thus strong versions of sustainable development represent a greater challenge to the economic and political *status quo* than do weak versions. Strong versions of sustainable development also seek much greater levels of democratic participation in processes of economic, social and environmental decision-making than do weak versions, which instead tend to emphasise the importance of 'top-down', state-led initiatives. In their strongest forms, versions of sustainable development represent profound visions of alternative societies operating by radically different economic, social and environmental principles (such as the principles of ecocentrism and deep ecology) and those versions may in fact reject notions of 'development' altogether. In their weakest forms, versions of sustainable development amount to little more than business-as-usual practices and minimal forms of pollution control. Needless to say, to date, many approaches to the management of environmental impacts have been based on weak, or very weak, versions of sustainable development.

Despite the proliferation of meanings of sustainable development, the lack of consensus about the validity of the concept, and the important differences between different versions of sustainable development, the term has entered the 'mainstream' of development theory, policy and practice. Debates about economic growth, about social development and about the management of environmental impacts are now often framed within broader debates about promoting sustainable development. Whilst the concept remains problematic, contested and ambiguous, it has nonetheless acquired some core principles (Dresner 2008). The most basic principle is that sustainable development should not destroy the natural resource base upon which human societies depend for their livelihoods (Ponting 2007). Development depends upon the natural environment that provides resources and assimilates pollution; development is clearly 'unsustainable' – insofar as it cannot continue indefinitely – if economic growth exhausts the available supplies of raw materials, the sources of energy or the pollution sinks. If excessive demands are placed upon the natural resource base (through deliberate or inadvertent exploitation or mismanagement of the environment), then both natural processes and human activities are bound to decline – perhaps with catastrophic consequences. The over-exploitation of natural resources, particularly in situations where the warning signs and consequences of environmental degradation were disregarded, has in the past led to the collapse of social and ecological systems (Ponting 2007). Hence sustainable development requires the maintenance of a sufficient resource base to allow survival; that task requires that economic, social and environmental considerations are balanced in making decisions about the benefits and costs of development. Whilst notions of limits to growth are unpopular, many definitions of sustainable development include the idea that the environment presents fundamental limits to human activities.

In addition to the (commonsense) notion that human societies should not destroy their life-support system, sustainable development includes two other vital principles: those of intergenerational equity and intragenerational equity. The principle of intergenerational equity suggests that humans must preserve a sufficient natural resource base, not simply for the current human population, but also for future generations of people. Failure to do so would clearly be unsustainable because it would create the conditions in which future generations would be left without the resources necessary to maintain an equivalent way of life, a situation that would most likely result in conflict, social deterioration, accelerating environmental degradation and eventually social and ecological collapse. In contrast, the principle of intragenerational equity suggests that the wide disparities in levels of affluence, opportunity, wellbeing and quality of life existing within each generation must be minimised. Development cannot be characterised as 'sustainable' if it allows the persistence of widespread poverty. The *Human Development Report 2005*, published by the United Nations Development Programme (UNDP 2006), indicated that, despite decades of efforts aimed at bringing about development, poverty remains an immense challenge worldwide. Around 2.8 billion people live on less than US$2 a day; those people lack access to education and healthcare and they are vulnerable to illness, violence, natural disasters and a wide range of other problems. This is not to deny that some progress in promoting poverty reduction has occurred. The proportion of people in the world living on less than US$1 a day halved during the period 1981–2001, and the absolute number of people living at that level began to decline. Nevertheless, the distribution of those improvements has been highly uneven and most improvements have been achieved in China. If China is excluded from the analysis, the number of people living on US$1 a day has in fact increased, especially in Africa. Overall, the brutal reality is that more than a fifth of the world's population still lives in extreme poverty. International development targets, such as the 2015 Millennium Development Goals (MDGs), have been agreed in an attempt to drive substantial progress in poverty reduction. Yet doubts exist about whether those targets will be achieved; and, even if they are, around 900 million people will remain living in chronic poverty. The *World Development Report 2009: Reshaping Economic Geography*, published by the World Bank (2009), confirmed that stark contrasts persist at the global scale between the rich and the poor, and that observation was reiterated – in the context of the increasing threat posed by climate change – in the *World Development Report 2010: Development and Climate Change* (World Bank 2010). Poverty reduction is therefore a central concern of sustainable development. Thus both of these principles – intergenerational and intragenerational equity – are essential elements in ideas of sustainable development.

In addition to the principles mentioned above, the task of promoting sustainable development involves the acceptance of a range of normative principles about good governance (Baker 2006). That observation follows from insights that the process of development is ideally concerned with bringing about social change that allows people to achieve their human potential, so is ultimately about the enhancement of individual freedoms (Adams 2009; Sen 1999). Some commentators have argued that such social change requires fundamental changes in modes of governance in order to promote the common good and to steer society towards collective goals (Baker 2006; Dresner 2008). The apparent failure of traditional forms of government intervention and policymaking to curb the escalating problem of environmental degradation and the persistence of poverty has prompted many people to question whether alternative modes of governance are required in order to promote sustainable development. Baker (2006, 9) has stated: 'Rather than being the task of national governments acting alone and using traditional policy means, promoting sustainable development requires engagement across all levels of social organization, from the international, national, sub-national, societal to the level of the individual.' Such comments reflect the fact that some of the most pressing environmental issues (such as biodiversity loss and climate change) are global in their scope and require transboundary, co-ordinated action – whilst simultaneously much development planning must be responsive to sub-national, regional and local variations in needs, concerns and capacity. Consequently, many 'new forms of environmental governance' have emerged: they include the participation of non-state actors, together with state and international organisations, in decision-making; they also involve the use of a wide range of policy instruments – including regulatory, market-based and voluntary instruments – together with various normative governance principles to promote sustainable development (Baker 2006). Considerations of participation and justice are prominent in debates about the nature and use of such new forms of governance.

Ideas about sustainable development incorporate various other principles, some of which have emerged in debates about environmental management: two of the most prominent are the precautionary principle and the polluter-pays principle. The precautionary principle suggests that, where there are threats of serious or irreversible damage to human health or to the environment, the lack of full scientific certainty should not be used as a reason for postponing action; this principle is especially relevant to anthropogenic impacts on the atmosphere because those impacts may be difficult to detect and may involve highly complex feedback mechanisms that are poorly understood (Houghton 2009; Roberts 2004). Thus this principle suggests that the management of human activities should err on the side of caution if uncertainties exist about their environmental or social impacts, even if such a course of action means forgoing short-term economic gain. The precautionary principle was articulated in the 1992 Rio Declaration on Environment and Development (Principle 15) and it now forms a central element of sustainable development frameworks. As Baker (2006) has pointed out, strong versions of sustainable development emphasise the significance of the precautionary principle to a much greater extent than do weak versions; hence, in stronger versions of sustainable development, market forces should not be allowed free rein to determine human activities, but new patterns of sustainable production and consumption should instead be encouraged through a combination of strong state intervention (government) and new forms of participation (governance). The polluter-pays principle suggests that any party causing environmental degradation should bear the full cost of that degradation: either the cost of environmental remediation or the cost in terms of lost environmental goods and services. Again, strong and weak versions of sustainable development differ in the rigidity which this principle is applied. Strong versions of sustainable development tend to require polluters to make strict reparation for environmental degradation (in the form of the remediation of damaged habitats or the creation of equivalent environmental capacity elsewhere), whilst weak versions of sustainable development are more likely to require polluters to compensate for lost natural capital by the creation of equivalent human capital.

One further, important aspect of sustainable development is that, given its breadth of concerns and its multi-scale, multi-sector scope, it cannot simply be assigned a discrete place within the activities of organisations; instead, it is a philosophy that should ideally be deeply embedded in economic and social structures – and that should permeate the activities of organisations and individuals. Thus the promotion of sustainable development is increasingly acknowledged to be a cross-cutting issue that affects all economic, social and environmental activities. Such an insight is reflected, for instance, in the UK sustainable development strategy, *Securing the Future* (UK Government 2005). One implication of this insight is that the promotion of sustainable development requires policy integration across all other policy areas; yet, currently, many inconsistencies and contradictions exist between policies, especially between economic development, energy, transport and environment policies. Achieving such policy integration represents a considerable challenge for policymakers, who must attempt to balance a wide range of differing – and often conflicting – economic, social and environmental concerns and demands. Yet, above all, the need to promote sustainable development presents a profound challenge to conceptions of development that prioritise individual self-advancement; instead, sustainable development seeks to place the common good before the over-exploitation of common resources by an affluent and powerful minority (Baker 2006). As Baker (2006, 215) has argued:

> the challenge to promote sustainable development is not just about finding more effective and efficient institutions of environmental governance. It is also about genuine commitment to a common interest, developing new ecologically and socially based values and focusing on human rather than state security. It is ultimately about the distribution of power, between the global and the local, between the privileged and the marginalized, and about the priority given to the economic, the social and the environmental, at present and in the future.

Ultimately, therefore, sustainable development is not about the way in which environmental impacts are managed but about who has the power to decide how those impacts are managed. Sustainable development is 'a way of talking about the future shape of the world': the beginning of a process of political reflection and action, a statement of intent and a challenge to action (Adams 2009, 379).

The Relationship between Air Transport and Sustainable Development

The concept of sustainable development is complex and multi-faceted: it may be interpreted in many ways – and this can lead to problems when trying to assess whether or not specific human activities contribute to the promotion of sustainable development. (It is worth bearing in mind that, whilst some activities may be clearly contrary to the promotion of sustainable development, the concept itself refers to the broader, societal, level and may be even more problematic when applied to particular economic sectors or activities in isolation.) Yet, for all the slipperiness of the concept, sustainable development involves certain core principles: balancing the economic, social and environmental benefits and costs of activities; protecting the natural resource base upon which human (and other species') survival depends; ensuring intergenerational and intragenerational equity, including a strong commitment to poverty reduction; and seeking more participatory and just forms of government and governance. One way of assessing the relationship between air transport and sustainable development is to ask to what extent the operation of the air transport industry is aligned with those core principles. The broad popularity of ideas about sustainable development means that the air transport industry, like many other activities, is under considerable pressure to demonstrate its commitment to promoting sustainable development. As a result, there has been rapid growth in the rhetoric of air transport and sustainable development during the last decade. In this section, some of that rhetoric is examined. Most of the pronouncements of national and international authorities, and of the air transport industry itself, are based on weak or very weak versions of sustainable development, with the result that those statements may amount to little more than commitments to business-as-usual – or even to stronger aviation growth – albeit with a greenwash of phrases about environmental responsibility and sustainability. Unsurprisingly, many authors have argued that recent trends in air transport are actually far from being 'sustainable' and that the projected growth of aviation is incompatible with the requirements of sustainable development, especially if stronger versions of sustainable development are envisioned. Furthermore, the idea of sustainable development is often used to justify the environmental impacts of air transport on the grounds that those impacts are far outweighed by the economic (and associated social) benefits of the industry. Yet, on closer examination, the economic benefits of aviation may not be so straightforward – which raises doubts that those benefits outweigh the environmental impacts of the industry.

The Rhetoric of Sustainable Development

The rhetoric of sustainable development frequently appears in debates about aviation environmental impacts; indeed, those debates often involve the use of terms such as 'sustainable aviation', 'sustainable mobility', 'sustainable society', 'sustainable change' and – even more problematically – 'sustainable growth' (ATAG 2009; Sustainable Aviation 2005; Upham et al. 2003). In a paper prepared in 2001 for the Ninth Session of the Commission on Sustainable Development of the United Nations Department of Economic and Social Affairs (DESA), entitled *Aviation and Sustainable Development*, the International Civil Aviation Organization (ICAO) acknowledged a now-familiar suite of public concerns: the rapid growth of demand for air transport; the main environmental impacts of the air transport industry; and the effects of a range of possible mitigation measures, including various technological and operational measures and policy instruments (ICAO 2001). However, that document scarcely mentioned sustainable development and it reflected nothing of the concept or its associated principles. The following year, in its *Statement by the International Civil Aviation Organization (ICAO) to the World Summit on Sustainable Development*, ICAO (2002, 1) hinted at the need to balance the economic, social and environmental costs and benefits of aviation, although the remainder of that document consisted of well-rehearsed statements about the economic and social benefits of the air transport industry and the need for states to support ICAO's 'common, harmonized approach' to achieving incremental improvements in the environmental performance of the industry. Since that year, ICAO has demonstrated little progress in developing a fuller understanding of – or in actually engaging with – the concept of sustainable development. A more nuanced understanding of the principles of sustainable development is apparent in some EU documents: for instance, in the communication entitled *Air Transport and the Environment: Towards Meeting the Challenges of Sustainable Development* (CEC 1999), in which the unsustainable nature of recent trends in aviation, the need for the air transport industry to compensate for the full environmental effects of its growth and the need for greater policy integration were acknowledged. Again, however, that document stops short of engaging with the more radical requirements of sustainable development and instead focuses on a range of possible measures for the management of aviation environmental impacts.

At the national level, pronouncements about air transport and sustainable development tend to be linked with the national sustainable development frameworks produced as a result of the 'UNCED process': the range of national and international activities that have occurred following the publication of the WCED (1987) report, including the international agreements made at the Rio and Johannesburg summits, held in 1992 and 2002, respectively (Baker 2006). The UK Department for Transport (DfT), for example, has stated that an 'overall policy aim of sustainability in aviation is being sought in the context of the UK's commitment to constructive engagement in ICAO and the EU', and that the UK Government 'is committed to ensuring that the long term development of aviation is sustainable. This means striking a balance between the social, economic and environmental aspects of air transport' (DfT 2002; 2003, 6). The UK Civil Aviation Authority (CAA), in its statement entitled *CAA Sustainable Development and Aviation Environmental Policy*, has acknowledged the need to 'enable civil aviation to best meet the needs of its users and society in a safe and sustainable manner'; in that policy document, the CAA accepted the UK Government's definition of sustainability as aiming to ensure 'a better quality of life for everyone, now and for generations to come' (CAA 2008, 1; UK Government 2005). That view of sustainable development involves four main objectives: the maintenance of high and stable levels of economic growth and employment; the achievement of social progress which recognises the needs of everyone; the prudent use of natural resources; and the effective protection of the environment (CAA 2008; UK Government 2005). Once again, such a formulation represents a weak version of sustainable development that emphasises the need for economic growth and its associated social benefits in conjunction with the more effective management of the resulting environmental impacts (especially by technological means). Underlying such a formulation of sustainable development are various assumptions about the fundamental legitimacy of converting natural capital into human capital – and about the capacity of humans to fully understand and mitigate environmental impacts.

Other national-level statements about air transport and sustainable development have been produced, including several documents by the UK Sustainable Development Commission (SDC) and the UK-based Institute for Public Policy Research (IPPR). Adopting a relatively weak version of sustainable development, the SDC (2001a) stated that 'sustainable development is about recognising as far as possible the legitimacy of people's aspirations, and finding ways of meeting them which will not prejudice the interests of other people, now and in the future.' The SDC (2001a) acknowledged that aviation has been 'hugely wealth creating and liberating for nations, regions and individuals' and that 'aviation has been a long-term success story for the UK economy'. Nevertheless, the SDC (2001a) called for absolute environmental limits to be applied to the air transport industry 'so that the aviation industry does not drive and contribute to the creation of an unsustainable economy and society'. In a subsequent publication, the SDC (2002) outlined six fundamental principles of sustainable development: (a) putting sustainable development at the centre; (b) valuing nature; (c) fair shares; (d) polluter pays; (e) good governance; and (f) adopting a precautionary approach. In that document, the SDC (2002) adopted a much stronger approach to sustainable development, arguing that the UK Government's air transport policy proposals fell 'seriously short of sustainability in respect of all these basic principles' and that they avoided 'the much deeper analysis which a truly sustainable approach would require'. Consequently, the SDC (2002) concluded, the rate of aviation growth had already become unsustainable and 'policy should be seeking to manage growth rates towards more sustainable levels'. The SDC (2002) warned that 'present trends in the growth of air traffic are leading the economy to an excessive and dangerous dependence on air travel and the resources that it consumes'; thus the SDC (2002) called for 'a more radical rethink' of air transport policy along lines that are more consistent with the principles of sustainable development (see SDC 2008).

The relationship between air transport and sustainable development has also been considered by the IPPR (Bishop and Grayling 2003; Hewett and Foley 2000). In an influential document entitled *The Sky's the Limit: Policies for Sustainable Aviation*, Bishop and Grayling (2003, 13) argued that 'economic growth is the engine of sustainable development' – although they acknowledged that the quality of that growth is an important consideration. Bishop and Grayling (2003) also acknowledged that there exist fundamental environmental limits to human activity, and that both technological and social (lifestyle) changes are required in order to allow economic growth within those environmental limits; thus their approach is much closer to the original WCED (1987) formulation of sustainable development than that of many other authors. Bishop and Grayling (2003) demonstrated that two different versions of sustainability underlie different attitudes to the environmental impacts of air transport: ideas of 'operational sustainability' (which focus on the ability of the air transport industry to meet future demand and to generate maximum economic growth and employment); and an alternative, environmentalist version of sustainability (which regards air travel as being representative of a fundamentally unsustainable pattern of consumption and which seeks ways to constrain air transport as much as possible in order to minimise its environmental impacts). Bishop and Grayling (2003) argued that both of those versions fail to adopt a sufficiently broad perspective, since sustainable development requires that both the economic benefits (and the associated social benefits) and the environmental impacts of air transport should be balanced. Specifically, Bishop and Grayling (2003) argued that the contribution of the air transport industry to various sustainability objectives – defined as the economic, social and environmental benefits and costs of projected aviation growth – should be used as the criteria for making a balanced judgement about air transport policy. Such a view reflects a relatively weak version of sustainability, however, since it implies that the loss of natural capital may be compensated for by equivalent increases in human capital.

If the rhetoric used by national and international authorities generally reflects weak versions of sustainable development, that used by the air transport industry is weaker still – and in many cases is simply a greenwash painted over business-as-usual practices. The fact that the air transport industry is highly dependent upon kerosene supplies and is strongly affected by fuel prices means that a constant commercial incentive exists for airlines to reduce their fuel costs by improving fuel efficiency, which has the desirable side-effect of reducing most aircraft emissions. The incentive to improve fuel efficiency is, in turn, passed on to airframe and engine manufacturers, since more fuel-efficient aircraft are constantly sought by airlines. In addition, airport operators and air traffic management (ATM) service providers have strong commercial incentives to improve the efficiency of their operations, which may also result in reduced emissions. All of the major actors within the air transport industry therefore have constant commercial incentives to improve their operational efficiency; the fact that, during the course of their normal business practices, those actors may incrementally reduce their emissions falls a long way short of promoting sustainable development. Nevertheless, some representatives of the air transport industry have been quick to claim that their practices represent a genuine commitment to 'sustainability'. The 'global approach to sustainability' outlined by the Air Transport Action Group (ATAG) includes the call 'to maximise air transport's growth potential in line with market demand, by pressing for the expansion and improvement of airports, air route capabilities, ground access, etc. according to sound environmental criteria' (ATAG 2009). Optimistically, ATAG (2009) argued that the environmental effects of such growth are of no cause for concern because 'air transport has been able to reduce or contain its environmental impact by continually improving its fuel consumption, reducing noise and introducing new, more sustainable technologies' – although the nature of those 'new, more sustainable technologies' is not specified. In its statement, *Industry as a Partner for Sustainable Development*, ATAG (2002, 5) emphasised 'air transport's full commitment to sustainability': a commitment that involves 'defining the right balance between economic, social and environmental mobility factors for our society'. ATAG (2009, 5, 7–8) also argued that strategic partnerships are needed 'to meet society's growing mobility needs as a direct response to the major challenges facing air transport in a sustainable world'. According to ATAG (2002, 31): 'Sustainable development is about ensuring a better quality of life, now and for generations to come. [...] It offers the air transport industry, and other businesses, opportunities to meet worldwide demands in a sustainable way.'

Better progress in engaging with the principles of sustainable development has been made through the Sustainable Aviation partnership, which emerged as an initiative of the UK Department of Trade and Industry (DTI) and the Department for Environment, Food and Rural Affairs (DEFRA). In particular, the partnership's high-level statement, entitled *A Strategy Towards Sustainable Development of UK Aviation*, presented 'the strategy developed by the UK aviation industry to respond to the challenge of building a sustainable future' (Sustainable Aviation 2005, 6). In that document, Sustainable Aviation (2005) acknowledged that the future growth of the air transport industry represents a major challenge to the sustainable development of the industry and that more effective management of aviation environmental impacts is required. Yet Sustainable Aviation (2005, 6) argued that the air transport industry faces considerable challenges in aligning its practices with the principles of sustainable development; for aviation, 'the context of sustainable development must include acknowledgement of some key features, for example the long lead-times for the implementation of new technology and the lack of alternatives to fossil fuel.' Overall, the Sustainable Aviation (2005, 9) document focused on the role of 'aviation in a sustainable society': a subtle departure from the (perhaps less comfortable) idea of 'sustainable aviation'. Nevertheless, the Sustainable Aviation (2005, 8) document set out a wide range of issues, commitments and recommendations for action, and it acknowledged that the strategy represented the first step in an iterative process of promoting sustainable development across the international aviation sector. Whilst the Sustainable Aviation strategy, unsurprisingly, adopts a weak version of sustainable development, it nevertheless represents a constructive, dialogue-promoting contribution to recent debates about air transport and sustainable development.

The Economic Benefits of Air Transport

As sustainable development involves a balancing act between the economic, social and environmental benefits and costs of human activities, the concept is frequently used as a rationale for justifying the environmental impacts of air transport on the grounds that the economic benefits (and the associated social benefits) of the industry far outweigh those impacts. It is worth exploring that benign view, analysing the assumptions upon which it is based, because the relationship between air transport and its economic benefits may be more complex than is commonly held. Several influential studies have shown that air transport makes a significant contribution to economic development, to the extent that aviation is described as a major 'engine' of economic growth (Boon and Wit 2005; DfT 2003a; OEF 1999; 2002; 2006). In their assessment of various studies of the economic contribution of aviation, Boon and Wit (2005, 1) acknowledged that the continued growth of environmental impacts in the vicinity of European airports has generally been justified 'largely on economic grounds'. In another study, focusing on policies for sustainable aviation, Bishop and Grayling (2003, 5) stated that there are 'sizeable economic and social benefits associated with air transport' including the contributions to economic prosperity made by business travel and air freight operations. Overall, Bishop and Grayling (2003) summarised the economic benefits of aviation in terms of access to markets, specialisation, economies of scale and foreign direct investment (FDI), and the social benefits in terms of employment, leisure travel, cultural exchange, consumer choice and opportunities to visit family and friends. The economic benefits of air transport have been investigated in considerable detail for one country, the UK, and a review of those studies raises some important questions about the nature of those economic benefits and the ways in which air transport could potentially be used to promote sustainable development.

In an authoritative analysis of the contribution of aviation to the UK economy, Oxford Economic Forecasting (OEF) argued that the air transport industry makes a substantial economic contribution to the UK – primarily through its impact on the performance of other industries and through supporting their growth. Since the UK is an island nation located at the edge of Europe, the study argued that the economic contribution of its aviation industry is probably larger than those of other countries (OEF 1999). Yet the study found that aviation also makes a substantial contribution to the UK economy in its own right in the following ways: (a) by adding £10.2 billion to GDP, 1.4 per cent of the total; (b) by directly employing 180,000 people, 0.8 per cent of the total; (c) by supporting up to 540,000 additional jobs indirectly through the supply chain, induced effects and jobs depending upon inbound and outbound travellers; (d) by promoting high productivity, generating around 2.5 times as much value-added per capita as the average UK industry; (e) by exporting £6.6 billion of services, 11 per cent of UK service exports and 3 per cent of total UK exports; (f) by transporting a further £35 billion of goods, 20 per cent of total UK goods exports; (g) by contributing at least £2.5 billion in taxes; and (h) by investing £2.5 billion during the period 1995–1999, 3 per cent of total UK business investment (OEF 1999). Thus the OEF (1999, 5) study argued that aviation is a key component of the UK transport infrastructure on which other parts of the economy depend, and that 'investments in that infrastructure boost productivity growth across the rest of the economy'. The study found that this occurs in several ways. First, better transport links can expand markets, which in turn allow greater economies of scale, increased specialisation in areas of comparative advantage, and stronger competitive pressures upon companies to achieve greater efficiencies. Air transport in particular allows those markets to be expanded to the global scale, supporting FDI both into and out of the UK – frequently accompanied by technological improvements. Second, improved transport links stimulate innovation, due to more effective networking and collaboration over longer distances (OEF 1999). Aviation thus boosts the rate of economic growth for several reasons: (a) because it is a rapidly growing sector in its own right, increasing at four times the rate of the UK economy as a whole during the period 1990–1999; (b) because the economic sectors that are most likely to support future economic growth are the same ones that depend heavily upon aviation; and (c) because efficient transport links are a key factor influencing where international companies choose to invest, and good air transport provision is believed to be critical to attracting inward investment in high technology sectors (including electronics and life sciences) and in key functions (including head office and research and development). Consequently, the UK economy as a whole is projected to become more dependent upon aviation in the future (OEF 1999).

In an updated study of the contribution of the aviation industry to the UK economy, OEF (2006) identified the following key aspects of that contribution: (a) in 2004, aviation directly added £11.4 billion to UK GDP, 1.1 per cent of the total; (b) in the same year, aviation directly employed 186,000 people in the UK and over 520,000 UK jobs in total depended upon aviation; (c) visitors arriving by air add over £12 billion annually to the UK tourism industry, generating an additional 170,000 jobs; (d) 55 per cent by value of UK manufactured exports to countries outside the EU are transported by air; (e) air services are particularly important for UK trade with fast-growing emerging economies, such as China, and for trade in high-value goods and services; (f) air services are vital for the growth sectors on which the future economic success of the UK is believed to depend, such as high-tech industries and financial and business services; (g) air services improve the competitiveness of almost all aspects of companies' operations; (h) air services expand markets and stimulate innovation and efficiency; (i) one quarter of all UK companies report that access to air services is important in determining their choice of location; (j) increased business use of air services would probably generate substantial wider economic benefits from improvements in productivity throughout the economy, which could generate additional GDP of over £13 billion annually (in 2006 prices) by 2030 with a net present value of £81 billion; (k) congestion costs have been rising more rapidly than air transport capacity, with the result that congestion costs to airlines and passengers were estimated to be £1.7 billion in 2005 and could exceed £5 billion (in 2006 prices) by 2015 if recent trends continue; and (l) the economic benefits of providing additional air transport infrastructure remain considerable even after the climate change costs of additional emissions are taken into account (OEF 2006).

In an assessment of the regional economic benefits of UK aviation, OEF (2002) argued that the benefits of direct employment due to aviation did not occur only in the regions where the jobs are located, but also extended to other regions. The study claimed that indirect employment was more evenly distributed across the UK than direct employment, because large airports support extensive supply chains that may reach other parts of the country. However, OEF (2002) acknowledged that substantial regional differences exist in both the rates of productivity growth and the contribution of aviation to GDP. Those regions with the most extensive air transport links – both within the UK and with other parts of the world – were projected to display the strongest effects on regional economic growth through improved productivity, enlarging markets and encouraging innovation, competition and economies of scale (OEF 2002). Another investigation of the contribution of aviation to regional economic development was commissioned by the DfT (2003a); that study emphasised the potential role of air transport in supporting connectivity, competitiveness and the growth of core cities. Specifically, the DfT (2003a) study acknowledged that aviation provides direct and indirect employment, travel time savings and broader catalytic effects; those broader catalytic effects include the advantages of attracting international headquarters to regional locations, the benefits of export and import trade, inbound tourism, knowledge transfer and positive effects on investment decisions (DfT 2003a). Overall, the studies mentioned above all argue that air transport makes a strong, positive contribution to economic development. Whilst that contribution is made at the cost of considerable environmental impacts, and with mixed social impacts, it is nevertheless judged to far outweigh the environmental and social costs of aviation (OEF 2006).

In this chapter, the intention is neither to dispute the empirical basis of such claims nor to evaluate the precise economic, environmental and social costs and benefits of air transport; those tasks would require another book. Instead, the purpose here is to question some of the assumptions that are implicit in the conclusions of those authoritative studies. In particular, the studies mentioned above face an entirely different type of criticism: they imply that the relationship between air transport and economic development is straightforward and that it will invariably deliver strong, positive benefits that are relatively evenly distributed throughout economies and places. Even accepting those authoritative claims about the magnitude and benign nature of the economic benefits offered by air transport, the question of whether such benefits extend to all parts of the global economy deserves more critical attention. Therefore, it is important to explore those assumptions and to ask critical questions about the applicability of those claims to all sectors and to all geographical regions. Many assumptions are implicit either in the studies mentioned above or in policy statements that are based on their findings (ATAG 2002; DfT 2003b; Sustainable Aviation 2005). In the absence of indications to the contrary, the economic benefits of air transport are assumed in those authoritative studies and policy statements to be: (a) delivered across the whole national economy; (b) delivered across all economic sectors; (c) delivered internationally (and, potentially, globally); (d) spatially uniformly distributed; (e) temporally uniformly distributed (or, alternatively, increasing over time); (f) uniform in terms of their quality; (g) neutral in terms of power or dependency between regions or nations; (h) reciprocal (whether bilateral or multilateral); (i) enduring and secure as a basis for investment and social development; (j) threatened by insufficient air transport capacity; and (k) liable to increase in proportion with air transport capacity. In turn, these assumptions are based on further assumptions, such as the assumption that national and regional economies function in similar ways worldwide, and the assumption that increased air transport capacity is always desirable, wherever it is provided.

In fact, critical analysis suggests that such assumptions are at best untested, at worst fallacies. Many studies have confirmed that the economic benefits of air transport are concentrated in core regions of national economies that are already intensively developed with transport infrastructure, especially where major hub airports benefit from large numbers of transit passengers and are thereby able to service a wider range of destinations and a greater frequency of flights than could be supported by local demand alone (Bishop and Grayling 2003; DfT 2003b; Graham and Guyer 1999; Morrell and Lu 2006). Economic benefits of air transport do not accrue across the whole of a national economy; indeed, specific, targeted government intervention is necessary if regional imbalances in demand for, and availability of, air transport are not to become self-perpetuating (Bishop and Grayling 2003; Gössling et al. 2009; Jin *et al* 2004; Wang and Jin 2007). Nor are those economic benefits delivered uniformly across economic sectors; many businesses do not require air transport services, nor do they benefit from inbound or outbound passenger movements, yet they may nonetheless incur costs due to airport-related congestion. Even more significant conceptual problems are associated with assumptions that the economic benefits of air transport are delivered internationally or even globally. Civil aircraft movements are highly unevenly distributed, globally, with marked concentrations of air traffic in the North Atlantic, across North America and Europe, and increasingly in parts of Asia and the Pacific (Rogers et al. 2002; UNCTAD 1999a). In contrast, other parts of the world have few air transport services and limited air transport infrastructure. If the provision of air transport services is highly unequal, then it is likely that the benefits associated with aviation are far from uniformly distributed spatially.

Assumptions that air transport provision will continue into the future – or may increase with time – have received more critical attention from various commentators. Many authors have argued that the growth of air transport could potentially be curtailed by limited capacity, whether airport, airspace or environmental capacity (Graham and Guyer 1999; Ignaccolo 2000; Upham 2001; Upham et al. 2003). Here, the intention is not to evaluate the question of how capacity constraints might limit the growth of air transport, either globally or at individual airports, but rather to make the point that, even if air transport growth continues into the future, it will not do so in a uniform manner worldwide. On the contrary, the greatest increases in demand for air transport are projected to occur in areas that are already well serviced by aviation, especially the major hub airports and/or the high volume city-pair routes in North America, Europe, Asia and the Pacific (Airbus 2006; Boeing 2006; UNCTAD 1999a). Peripheral regions and nations in the world economy, and peripheral regions within countries, are much less likely to experience the rapidly increasing demand for air transport that characterises the highly connected world cities. In particular, demand for air transport in Latin America, the Middle East and Africa is expected to continue growing at a slower rate than elsewhere (UNCTAD 1999a). Forsyth (2007) has argued that the recent emergence of new airline business models and the introduction of new aircraft types mean that increasing use will be made of secondary (including regional) airports, with the result that some pressure will be relieved on the most severely capacity constrained airports. Yet even if some demand is serviced by secondary airports, in some countries, the distribution of air transport provision is still likely to remain highly unevenly distributed – both worldwide and within countries.

Given the recent emergence of the so-called low-cost carriers (LCCs) and their important role in current air transport markets, it is worth considering how LCCs could influence the distribution of the benefits of air transport. Graham and Shaw (2008, 1439) have acknowledged that the growth of air travel that has occurred in response to the emergence of LCCs is perceived to be advantageous for national and regional economic development. In principle, LCCs spread the benefits of air travel more widely by increasing accessibility to air transport services, particularly where they operate from secondary, regional airports (Grubesic and Zook 2007; Nilsson 2009). However, Dobruszkes (2006) has argued that the growth of LCCs is currently focused on European markets, reflecting the geography of EU air transport liberalisation. In addition, Dobruszkes (2006) stated that LCCs often compete with existing charter routes, and that they frequently negotiate exclusive routes linked to the option for secondary airports and/or niches. Another study has shown that LCCs have an important influence on air transport services in the United States (Grubesic and Zook 2007). Therefore, whilst LCCs can potentially broaden accessibility to air transport services and can complement the networks offered by full service network carriers, their benefits currently occur principally in liberalised European and US markets and focus on competitive routes that are already well-serviced by air transport. Furthermore, the cost-cutting measures used by LCCs (and in turn adopted by other carriers) may limit the economic and social benefits of their services: by depressing wages, through the extensive use of subcontracting, and by reducing multiplier effects (Dennis 2009; Gillen and Lall 2004). In addition, Graham and Shaw (2008) have argued that, because the low-cost model promotes the rapid growth of air travel whilst failing to account for its external costs, that model is incompatible with the principles of sustainable development.

Further assumptions about the economic benefits associated with air transport relate to the quality of those benefits: namely, that they are uniform in their quality; that they are neutral in terms of power or dependency between regions or nations; and that they are reciprocal – whether bilaterally or multilaterally. Thus, in the authoritative studies reviewed above, air transport is regarded as being essentially apolitical in that it potentially offers equivalent benefits wherever it occurs. A critical perspective towards these assumptions suggests that they are highly problematic, however. Air transport makes a particularly valuable contribution to economic sectors that depend on the rapid transport of high-value goods and services, including high-tech industries, and to those that rely on just-in-time delivery strategies, as OEF (2006) has acknowledged. In terms of inbound and outbound passenger travel, air transport can contribute significant value to tourism source areas and destinations (Gössling and Peeters 2007; Graham et al. 2008; May 2002; Mayor and Tol 2010). Furthermore, air transport is closely related to globalisation and has played an important role in facilitating the growth of multinational corporations and global financial and business networks (Goetz and Graham 2004; Hettne 2008; Janelle and Beuthe 1997; UNCTAD 1999a; 1999c; Young 1997). But the assumption that air transport makes an equally valuable contribution to all economic sectors is difficult to substantiate. Moreover, not all countries and regions are pursuing economic development paths based on industries that benefit significantly from aviation; without specifically targeted policy interventions, air transport is likely to play a much more modest role in those alternative economic development strategies.

The implicit view that air transport provision is neutral in terms of power and dependency also deserves critical scrutiny. UNCTAD (1999a, 3) has drawn attention to 'the disparity in competitive situations between the air carriers of many developing countries and those of most developed countries', both in past and future trends. UNCTAD (1999a) identified numerous issues faced by many developing countries in maintaining even basic, viable air transport services in a highly competitive international market. Consequently, UNCTAD (1999c, 4) reported 'the concern expressed by developing countries on the need to improve their participation in air transport markets on a level playing field'. Even where viable air routes can be maintained between developed and developing countries, it is unlikely that the economic benefits of those routes accrue equally to stakeholders in both groups of countries. Thus the economic benefits of air transport services between developed and developing countries are unlikely to be reciprocal, either bilaterally or multilaterally; more likely, those benefits will be obtained preferentially by airlines, airports and air navigation service providers based in developed countries, and the competitive advantage available to countries and regions with highly developed air transport systems will be reinforced.

Other assumptions relate to the potential of air transport to provide an enduring and secure basis for investment and social development. The aviation industry that is vital in supporting advanced economies may not automatically provide such development benefits to other economies. In reality, air transport markets – especially in the developing world – may be volatile and may provide little certainty to investors. The belief that the air transport industry and its economic benefits are threatened by insufficient air transport capacity and are liable to increase in proportion with new air transport capacity is, again, an assumption based on the experience of developed countries. Such a view may not apply to developing countries, in which capacity constraints – whilst being a consideration at many airports – do not present the acute business risk that they do at major airports in developed countries (UNCTAD 1999a). In turn, those views about capacity, investment and social development are based on a further set of assumptions, such as the assumption that national and regional economies function in similar ways worldwide, or the assumption that increased air transport capacity is always desirable wherever it is provided. In fact, capacity constraints may occur for sound social or environmental reasons that in turn help to conserve resources and to increase the profitability of local enterprises. In addition, in some developing countries, the effective control of capacity can serve a valuable function in ensuring higher fares and increasing the profitability of airlines (UNCTAD 1999a).

Even where economic benefits do increase with the greater provision of air transport services, therefore, they do not necessarily accrue to those most urgently in need of those benefits. Analysis of key studies of air transport and economic development – and of the implicit assumptions upon which they are based – thus raises critical questions about the potential for air transport to act as an effective tool for sustainable development. Who benefits from the provision of air transport services at local, regional, national, international and global scales? Who benefits from increased capacity in the air transport system, and who loses? When additional capacity is created or released, where does that occur and what factors affect decisions about the control of new capacity? Overall, does air transport bring about any reduction in the wealth gap between developed and developing countries? Does it facilitate the participation of developing countries in global air transport markets on an equitable basis? Does air transport deliver any measurable progress towards the achievement of international development goals such as the MDGs (DfID 2007)? How might air transport best promote sustainable development and greater equity and well-being (Holden and Linnerud 2007)? The ability of aviation to bridge gaps between countries and people – and to deliver important economic and social benefits widely – is frequently celebrated; yet the answers to such questions would probably suggest that the economic benefits of air transport are not as great as has been hitherto claimed, and that they are highly unevenly distributed, with developed countries – and especially their core regions – benefiting far more than other places. This has important implications for any assessment of the relationship between air transport and sustainable development: if the concept of sustainable development is being used to justify intensifying aviation environmental impacts on the grounds that the industry also provides important economic (and associated social) benefits, then it is vital to ensure that those economic benefits are not overstated.

Promoting Sustainable Development: the Role of Air Transport

In the section above, many questions have been raised about the relationship between air transport and sustainable development. In particular, analysis of the rhetoric found in some high-level statements about air transport and sustainable development raises questions about the extent to which stronger versions of sustainable development could potentially inform air transport (and broader) policy. Also, critical examination of some influential studies of the economic benefits of air transport suggests that those benefits may be overstated – and that the relationship between air transport and economic development may be more complex than is commonly assumed. In turn, those considerations point to other questions. What role does – or should – air transport play in promoting sustainable development? How could air transport contribute more effectively to promoting sustainable development? In particular, how might air transport make a greater contribution to one crucial aspect of sustainable development: poverty reduction? What would be the likely implications for the air transport industry and for society more generally? Crucially (reflecting the fact that sustainable development is ultimately a political concept), who should decide the direction of air transport (and broader) policy? To what extent could that policy be guided by new forms of governance and government that are more participatory and just? How might the transitions to such forms of governance and government be achieved? In this section, various aspects of the ways in which air transport might more effectively be used to promote sustainable development are discussed. In general, there are two main ways in which the air transport industry might more effectively promote sustainable development: (a) through minimising its own environmental impacts; and (b) by contributing to the achievement of broader sustainable development objectives. Some of the preceding chapters have indicated ways in which the main environmental impacts of air transport could potentially be reduced (see Chapters 3, 4 and 5). Whilst that task is urgent and essential, those mitigation options are not discussed again here; instead, the discussion below focuses on some of the main ways in which air transport might play a role in achieving broader sustainable development objectives.

Arguably the first step in improving the extent to which air transport contributes to promoting sustainable development is to acknowledge the unsustainability of the current situation. Despite the rhetoric found in some high-level statements by national and international authorities and by some representatives of the air transport industry, many authors have acknowledged that the projected growth of air transport – and perhaps even the current level of air transport activity – is clearly unsustainable (Åkerman 2005; Anderson et al. 2005; Bishop and Grayling 2003; Bows et al. 2005; 2006; 2009; Broderick 2009; Caves 1994a; 1994b; CEC 1999; Fawcett 2000; Gössling and Upham 2009; Graham and Guyer 1999; Lu 2009; May and Hill 2006; Nilsson 2009; Peeters et al. 2006; 2009; Price and Probert 1995; Randles and Mander 2009; RCEP 2002; Upham 2001; 2003; Upham and Gössling 2009; Upham et al. 2003; 2004). As Anderson et al. (2005, 50) have suggested, it is necessary to 'convert the rhetoric into reality' if air transport is to be consistent with the principles of sustainable development. The fact that considerable doubts exist about the extent to which current and projected trends in air transport are 'sustainable' points to the urgent need for a comprehensive sustainability assessment of the sector. Yet, at present, no generally-agreed methodology exists for undertaking a comprehensive sustainability assessment of a whole sector of the economy and its alternative future pathways (SDC 2002). Nevertheless, the SDC (2002) has argued that a comprehensive sustainability assessment for the air transport industry is vital; and such a process could also help to develop the methodology required for the assessment of other economic sectors. Such an assessment could also allow the rhetoric of air transport and sustainable development to be critically evaluated, and would provide the basis for more informed decision-making and policymaking about air transport and sustainable development.

The next step in promoting a closer alignment between air transport and sustainable development is to understand human needs for air transport with greater precision (SDC 2002). Further research is needed to investigate the range of needs that people express for air transport – and, importantly, the reasons underlying those expressions of need. There are two main aspects to this issue. First, people frequently use air transport as a means to an end, such as visiting family and friends, conducting business, pursuing educational opportunities, accessing far-flung tourism destinations and travelling for medical treatment. The SDC (2002) has argued that some of those needs could – and perhaps should – be met in other ways besides the use of air transport: for instance, through the wider use of communication technologies in conducting long-distance business transactions. Second, air travel is sometimes sought by people as a good in its own right rather than as a means to some other end (Adey et al. 2007; Shaw and Thomas 2006). Both of these aspects have contributed to the general increase in demand for air transport and to the emergence of a group of 'hypermobile' travellers who undertake very frequent business- and leisure-related travel, often over long distances (Gössling et al. 2009; Shaw and Thomas 2006). Hypermobile travellers may in fact comprise several distinct groups of travellers: health migrants and medical tourists; second-home commuters (often to remote parts of the world); short-break, long-distance travellers (undertaking 'break-neck breaks'); frequent low-fare holidaymakers; long-distance commuters; long-haul business travellers; 'gap year' travellers; and those undertaking multi-sector 'sabbaticals' involving participation in voluntary projects (Gössling et al. 2009). Yet relatively little is known about the behaviour of hypermobile travellers and further research is required to elucidate their travel patterns and motivations (Gössling et al. 2009). Gössling et al. (2009) have argued that hypermobile travellers represent the group(s) with the greatest potential to achieve substantial reductions in their environmental impacts; those authors have also pointed out that it is equally important to dissuade other groups of currently less-frequent travellers from developing carbon-intensive, hypermobile lifestyles.

Promoting a situation in which air transport contributes more effectively to sustainable development could also involve processes of envisioning and exploring alternative paths for the future of the industry – and of society more generally. As the SDC (2002) has suggested, despite understandable economic aspirations, it is important that future visions for the air transport industry – and for society – engage with the common desire for a better quality of life 'in the round'. Such envisioning and exploratory processes could potentially raise many difficult questions. What sort(s) of economy, society and environment do people want? What role does mobility play in a desirable future? What forms of transportation might facilitate that level of mobility? In that context, what type of air transport industry would be desirable? Of course, such processes would generate high-level, aspirational statements that may be relatively difficult to translate into policy and action, especially given the existence of a multitude of conflicting views about the nature, purpose and value of 'development' (sustainable or otherwise). Nevertheless, defining alternative visions of the future paths of air transport and society, notions of the role of mobility and ideas about what constitutes wellbeing are, ultimately, essential tasks in order for policymaking to be directed towards the promotion of stronger, more holistic versions of sustainable development. Those alternative paths could, in turn, be used to develop richer scenarios and to inform more sophisticated models of linked social-ecological systems. Yet, despite the diversity of visions and scenarios that could potentially emerge from such envisioning and exploratory processes, that diversity is likely to include common, core elements. One common element is likely to be the recognition that the task of achieving an overall improvement in wellbeing and quality of life needs to be decoupled from continued growth in the consumption of resources and the generation of pollution – especially greenhouse gases (SDC 2002). In relation to air transport, the implication is that policy should seek to decouple aviation growth from economic growth rather than to reinforce their interconnections. Another common element is likely to be poverty reduction, since current or projected trends in aviation growth can never be regarded as promoting sustainable development if they allow the persistence of – or worsen – poverty (see below).

Therefore, processes of envisioning and exploring alternative future paths for air transport, and for society more generally, could inform better, more inclusive decision-making and policymaking. However, some aspects of future policy in relation to air transport are already clearly apparent. Crucially, if the air transport industry is to play a part in promoting sustainable development, it must first internalise fully its external (environmental and social) costs, including the costs incurred in monitoring and auditing its environmental and social performance. As the SDC (2001a) has stated:

> the costs and benefits (both social and environmental) which are created by aviation do not arise at the same places, to the same people, and at the same time. We do not think that future generations should pay costs to deliver benefits to current generations, or that western citizens should travel at the expense of climatic impacts on poorer societies.

Again, therefore, issues of justice, participatory governance and poverty reduction are raised in acknowledging the need for the air transport industry to internalise its external costs. As the preceding chapters have suggested, however, considerable debate has occurred about precisely how those external costs should be internalised (see Chapters 3, 4 and 5). The principles of sustainable development suggest that considerable flexibility may be required in order to respond adequately to the various global and localised environmental impacts of air transport, as well as to safeguard its most important economic benefits (SDC 2001a). Thus the SDC (2001a) suggested that the localised impacts of air pollution and aircraft noise are more likely to be managed locally or regionally, with a strong role for local communities in reaching agreements with airport operators, subject to basic safeguards and within the context of broader regulations. At the same time, the SDC (2001a) argued that the issue of climate change requires co-ordinated action at the global scale, and that air transport should be brought fully within the international climate change management regime. Various tensions and challenges exist, therefore, in ensuring a consistent, co-ordinated approach to the management of aviation environmental impacts whilst allowing sufficient flexibility to respond to regional and local concerns.

Ultimately, discussions of air transport and sustainable development cannot avoid the thorny issue of demand management, despite its unpopularity within the air transport industry and with national and international authorities (Anderson et al. 2005; Macintosh and Wallace 2008; Price and Probert 1995; Smith 2008). As Smith (2008) has acknowledged, rising transport demand is likely to be the biggest hurdle to reducing anthropogenic greenhouse gas emissions; and, whilst climate change is only one consideration within broader debates about sustainable development, it is one of the dominant global issues with the potential radically to alter human economies and societies and to negate the recent progress towards promoting human development (IPCC 2007; Stern 2006; UNDP 2007). Consequently, for the purposes of climate change mitigation alone, some authors have argued that demand management must inevitably become a feature of air transport policy. Anderson et al. (2005, 47) have argued that, for the UK, without government action to significant reduce aviation growth, 'the industry's emissions will outstrip the carbon reductions envisaged for all other sectors of the economy' – and that, if aviation growth exceeds just two-thirds of its current rate, even if efficiency improvements are taken into account and if all other sectors of the economy completely decarbonise, the UK Government's carbon reduction target will be impossible to achieve. Thus Anderson et al. (2005, 47) argued that 'it's not that we need to fly less, but that we cannot fly more'. For air transport, demand management would most likely include a package of measures. Price and Probert (1995, 157) argued that air travel demand management could include the following options: (a) substituting the greater use of telecommunications for air travel; (b) reducing the desire for long-distance leisure travel by improving local environments and leisure facilities; (c) avoiding needless air travel through better journey planning; and (d) increasing the cost of air travel by internalising the external costs imposed by aviation. However, those options are not equally palatable, feasible or equitable; as the SDC (2001a) has pointed out, 'reducing travel by reducing the need to travel is a far more sustainable solution than diminishing people's welfare just by pricing them out of a desirable activity.' Encouraging modal shift may be the most acceptable form of demand management. The SDC (2002) has acknowledged that, as has been the case for road traffic, air transport policy should seek to replace a simple 'predict and provide' approach with a more sophisticated mix involving appropriate pricing signals, intermodal shifts and the provision of new infrastructure.

Thus there are many, interrelated ways in which air transport could be aligned more closely with the principles of sustainable development. However, one important aspect of sustainable development – the notion of poverty reduction – has received relatively little attention to date in debates about the 'sustainability' of air transport. As the preceding section has suggested, debates about the future direction of air transport policy fall far short of engaging with the more radical requirements of sustainable development, including the urgent need for poverty reduction (Abeyratne 2003; 2004). Yet analysis of various United Nations (UN) documents indicates that air transport has a potentially crucial role to play in promoting sustainable development (UNCSD 2001; UNCTAD 1999a; 1999b; 1999c). The UN Commission on Social Development (UNCSD) acknowledged that transport services and systems (including air transport) should contribute to economic and social development as efficient and environmentally sound activities, and that they should be affordable and accessible in order to ensure mobility on an equitable basis to all sectors of society (UNCSD 2001). The benefits of air transport should be delivered in the areas of 'community health and safety, infrastructure, gender aspects, employment and labour conditions and providing for those with special needs' (UNCSD 2001, 3). Specifically, the UNCSD (2001, 3) stated:

> There is a strong need for adequate and efficient, economically viable, socially acceptable and environmentally sound transport systems, especially in developing countries where accessibility and affordability are important for the eradication of poverty, improving access to social services and access to employment opportunities. Prospects for achieving sustainable development depend on taking transport into account in urban and rural planning, public infrastructure decisions, and policies and measures to eradicate poverty and promote gender equality.

Consequently, the UNCSD (2001) urged governments to consider financing transport projects for sustainable development at the regional level, and to use appropriate measures to rationalise traffic flows, to manage transportation demand and to facilitate the flow of, and access to, goods. In particular, the UNCSD (2001, 7) acknowledged that the transport needs of the poor deserve special attention: access to transport services by poor people affects their ability to earn income as well as the cost of basic necessities and services needed for healthy living. Above all, transport planning 'should be guided by the principle of common but differentiated responsibilities' (UNCSD 2001, 7).

In a series of documents relating to air transport and developing countries, the UN Conference on Trade and Development (UNCTAD) also acknowledged the potential importance of air transport for developing countries, emphasising their need for greater and more equitable participation in air transport markets (Campling and Rosalie 2006; UNCTAD 1999a; 1999b; 1999c). UNCTAD (1999a) provided an overview of the multiple problems facing developing countries in accessing air transport markets, including the costliness of maintaining national flag carriers; the limited viability of domestic services; pressures to subsidise services to remote areas and communities; limited connectivity with major international air routes; small numbers of gateways per country; limited bargaining power in negotiating bilateral agreements; difficulties in gaining traffic rights; limited and outdated infrastructure; challenges in assuring standards of safety and quality; the vast gap between developed and developing countries in terms of marketing know-how and capabilities; scarce budgetary resources and small capital injections; and the aggressive, 'unfair competition' practices of some developed world carriers, including the use of 'fares below costs, the addition of excessive capacity or frequency of service, or the abuse of dominant position in a route' (UNCTAD 1999a, 14). As a result of such challenges, UNCTAD (1999a, 10) stated:

> The challenge is for developing countries to reconcile a number of potentially conflicting objectives [...]. Such conflicts generally derive from the fact that air transport plays a dual function for developing countries: it ensures external links to the major trading centres while at the same time providing for economic and social development, whether through its contribution to other economic sectors or through the provision of social services, for example to remote areas and communities.

Consequently, UNCTAD (1999a, 16; 1999c, 7) argued for 'a gradual, progressive, orderly and safeguarded change' towards increased access to international air transport markets by developing countries.

Air transport is thus acknowledged to be an important tool for economic and social development (Abeyratne 2003; 2004; Goldstein 2001; Miller and Clarke 2007; Raguraman 1995; Rhoades 2004). Yet the extent to which it actually serves that purpose is unclear, especially given that relevant, specific criteria, targets and performance indicators have not previously been defined. Below, various criteria are presented that could potentially be used to assess the extent to which air transport is an effective tool for sustainable development:

- What is the value of air transport's contribution to GDP in developing countries?
- What contribution does air transport make to direct and indirect employment in developing countries?
- What is the value of any multiplier effects due to air transport in developing countries?
- What measurable direct and indirect social benefits does air transport deliver in developing countries?
- What measurable direct and indirect social and environmental costs does air transport impose in developing countries?
- What measurable contributions does air transport make towards achieving international development targets such as the MDGs?
- To what extent do airlines, airports and air navigation service providers in developing countries participate in international air transport markets?

- To what extent are airlines, airports and air navigation service providers in developing countries hindered by protectionism and 'unfair competition' practices?
- To what extent does government support for airlines, airports and air navigation service providers in developing countries divert resources from other development needs?
- What measurable impacts does air transport have on human wellbeing in developing countries?

Several of these criteria make use of some standard indicators of economic performance (GDP, direct and indirect employment, multiplier effects). However, the remaining criteria listed above represent an attempt to go beyond a standard economic approach and to consider broader sustainable development objectives and notions of human wellbeing (Daley 2009; Gough and McGregor 2007). Those criteria relate to the measurable, direct and indirect social benefits – and the social and environmental costs – that occur in developing countries as a result of air transport services. Specifically, they include some consideration of whether – and to what extent – air transport assists in achieving international development targets such as the MDGs. Other criteria relate to the degree to which air transport organisations in developing countries are able to participate in international markets – or, alternatively, are hindered by protectionism and 'unfair competition' practices (UNCTAD 1999a). Where financial assistance is provided to air transport organisations in developing countries, the extent to which those resources have been diverted from other development needs could serve as another criterion by which the overall contribution of air transport to sustainable development may be assessed.

The concept of wellbeing has become a central organising principle in studies of development, and various methods for conceptualising, measuring and evaluating wellbeing have been developed (Gough and McGregor 2007). Those methods could be used to investigate the contribution of air transport to sustainable development. To some extent, work in this area has already commenced, with several studies exploring the economic, environmental and social issues relating to the air freight of horticulture exports from some African countries to the UK (Edwards-Jones et al. 2008; Gibbon and Bolwig 2007; MacGregor 2006; MacGregor and Vorley 2006; Vega 2008; Wangler 2006; Williams 2007; Williams et al. 2006). Further research could extend these studies to investigate other aspects of the air transport industry, and other locations. For instance, in an innovative study of the relationship between air transport and sustainability in the largest state in Brazil, Amazonas, Fenley et al. (2007)

suggested that the provision of air transport infrastructure in Brazil might form a valuable element of a broader sustainable development strategy; this is because, in remote, densely-forested regions, aviation may represent a better transportation option than road-building (which tends to be accompanied by extensive deforestation). In general, the contribution of air transport to sustainable development could be theorised in a more coherent, integrated way that takes into account the range of economic, environmental and social benefits and costs associated with the industry. Above all, posing the question of the benefits of air transport in terms of the requirements of those most urgently in need of those benefits – the poor – allows conventional claims about the potential of the industry to drive sustainable development to be evaluated.

Summary

In this chapter, various aspects of the complex relationship between air transport and sustainable development have been considered. Debates about environmental issues – particularly the critical global environmental issues of biodiversity loss and climate change – and their management now tend to be framed within broader debates about sustainable development (Adams 2009). Sustainable development is a highly problematic, contested concept that has become the central organising principle in debates about how to balance environmental protection with economic and social development. Different versions of sustainable development exist, ranging from very weak versions (that demand little more than a business-as-usual approach or minimal forms of pollution control) to much stronger versions (that place strict limits on the degradation of natural capital, or that may even reject the pursuit of economic growth and development altogether). Yet, as it is most commonly interpreted, the idea of sustainable development has several recurring, core principles, especially the need to balance the economic, social and environmental costs and benefits of development, both for present and future generations – an idea that incorporates the important principles of intergenerational and intragenerational equity (Adams 2009; Baker 2006; Dresner 2008; Elliott 2006). Whilst the performance of such a balancing act is attractive in theory (not least because it legitimises the pursuit of both economic growth and environmental protection), in practice reconciling a wide range of conflicting economic, social and environmental concerns is far from straightforward. Sustainable development is therefore a political concept, one that raises difficult questions about governance, participation, power and justice (Jacobs 1995; Redclift 1984). Crucially, sustainable development necessarily requires the reduction of poverty, and human activities that promote economic growth whilst allowing the persistence of poverty cannot claim to be 'sustainable' by any standard.

It is unlikely that the representatives of the air transport industry – and national and international authorities – intended to commit themselves to such a radical concept when they rushed to adopt the rhetoric of sustainable development. Indeed, confused notions of sustainable development were readily used in debates about air transport because the concept appeared to hold out the promise of justifying the continuing, rapid growth of aviation on the grounds that the industry provides important economic and social benefits. Thus, at first glance, the desire for 'sustainable development' apparently legitimated the intensifying environmental impacts associated with aviation growth. In fact, the principles of sustainable development require far more of the air transport industry than simply the use of standard fuel-conservation techniques that result (as a side-effect) in reductions in most emissions and in noise – together with wishful statements that the growing environmental impacts of its activity will be offset by technological and operational advances. Used in the sense in which it was originally intended by the WCED, the idea of sustainable development demands both improvements in eco-efficiency and profound changes in social organisation (Baker 2006; WCED 1987). Thus, in its most authoritative form, sustainable development requires radical changes in the processes and institutions of government and governance – including genuine and equitable participation in decision-making and policymaking at all levels of society – in order to promote justice, wellbeing and quality of life for people worldwide. With respect to the relationship between air transport and sustainable development, a central question is whether the growth of air transport should be constrained in order to keep aviation environmental impacts within acceptable limits, or whether aviation growth should be allowed and its intensifying environmental impacts justified on other (economic and social) grounds. As noted above, that question is about values, attitudes and beliefs far more than it is about technologies and operating practices, important as those factors are. And whilst the rhetoric of sustainable development has, in the past, been used to justify aviation growth, stronger versions of sustainable development increasingly point to the need to curtail that growth – especially in an attempt to mitigate the worst effects of climate change.

Yet the fact that the idea of sustainable development is more radical than is commonly assumed presents some novel opportunities for the air transport industry, as well as major challenges. Poverty reduction is a central aspect of sustainable development – and it is an idea that is likely to have been far from the minds of representatives of the air transport industry when they adopted the rhetoric of 'sustainability'. Yet poverty reduction is also one area in which air transport could potentially make a substantial contribution to the task of promoting sustainable development, partly because of the global reach and unique regulation of the industry, and partly because air transport has the capacity to create direct, rapid, responsive links between producers and consumers and it could potentially provide producers in developing countries with unprecedented access to international markets. Indeed, air transport is acknowledged to have a crucial role to play in promoting sustainable development, through providing producers with greater access to markets whilst at the same time contributing to the performance of other economic sectors and facilitating the provision of social services – especially to remote areas and communities (UNCSD 2001; UNCTAD 1999a; 1999b; 1999c). Many obstacles stand in the way of such progress, not least the need to ensure that developing countries enjoy greater and more equitable participation in international air transport markets, unhindered by protectionism or by 'unfair competition' practices (Campling and Rosalie 2006; UNCTAD 1999a; 1999b; 1999c). Nevertheless, air transport is acknowledged to have considerable potential to act as an important tool for economic and social development (Abeyratne 2003; 2004; Goldstein 2001; Miller and Clarke 2007; Raguraman 1995; Rhoades 2004).

Nothing absolves the air transport industry, policymakers and regulators of their responsibilities to achieve a compromise between aviation growth (and economic growth more generally) and environmental protection that is acceptable to the majority of people. Yet the management of those environmental impacts – whilst essential – should not blight economic development where it is most needed. The potential of the air transport industry to contribute to a dramatic reduction in poverty worldwide is one of the most exciting possibilities to emerge from recent debates about the relationships between air transport, economic growth, environmental protection and sustainable development. In his authoritative text on sustainable development, Adams (2009, 379) stated:

[Sustainable development] is not about the way the environment is managed but about who has the power to decide how it is managed. Its focus must be the capacity of the poor to exist on their own terms. At its heart, therefore [sustainable development] involves not just a pursuit of new forms of economic accounting or ecological guidelines or new planning structures, but an attempt to redirect environmental and developmental change so as to maintain or enhance people's capacity to sustain their livelihoods and to direct their own engagements with nature. Escobar (2004) calls for 'dissenting imaginations' that can think beyond modernity and the regimes of the globalized economy and the exploitation of marginalized people and nature. 'Sustainable development' is a way of talking about the future shape of the world. To conceive of the future in these terms marks the beginning of a process of political reflection and action, not the end. To call for sustainable development is not to set out a blueprint for action but to issue a statement of intent and a challenge to action.

Such 'dissenting imaginations' are now needed in order to chart a course for the future of the air transport industry (Escobar 2004). The relationship between air transport and sustainable development is essentially about the envisioning and exploration of alternative future paths for aviation – and for society more generally – and about putting plans into action in order to achieve the most desirable outcome(s). Having an adequate understanding of sustainable development is critical to the success of those tasks. If excessively weak versions of sustainable development are advocated by powerful interest groups and are adopted by policymakers, then the air transport industry will ultimately offer people little besides a superficial greenwash and a growing contribution to climate change. If, on the other hand, stronger versions of sustainable development inform air transport (and broader) policy, aviation could potentially change the future shape of the world in a much more creative, innovative and constructive way.

7 Conclusion

A Multitude of Issues

In this book, a multitude of issues relating to the subject of air transport and the environment has been considered. Those issues range from the global to the local scale; they involve technical and operational as well as policy considerations; and they include both scientific and socioeconomic dimensions. Above all, however, those issues are inherently political. Debates about air transport and the environment generate strong controversy – especially given the emergence of climate change as a global environmental issue of unprecedented significance – and the formulation and implementation of air transport policy involves balancing a wide range of (often conflicting) economic, social and environmental considerations. The provision of air transport services is an issue of international, national, regional and local importance involving powerful alliances, corporations and regulators. Over more than four decades, however, the emergence and spread of environmentalist concerns have led to the growth of other interest groups and movements; subsequently, strongly conflicting, entrenched positions have been adopted by both the pro-aviation and the environmentalist lobbies. To date, in general, air transport policy has been characterised by the attempt to promote the substantial growth of the industry whilst simultaneously attempting to mitigate the most objectionable environmental impacts using a relatively limited range of regulatory, market-based and voluntary instruments. The task of making effective policy for air transport is complicated by several factors: the strong, sustained growth of demand for air transport; the importance of the air transport industry in driving economic growth in other sectors; the strong links between aviation growth, tourism growth and processes of globalisation; the broad popularity of air travel; the fact that the industry is international in its scope, is regulated by myriad bilateral air service agreements (ASAs) and connects places with varying capacity to comply with environmental regulations; and, not least, the considerable uncertainties that remain with regard to some of the environmental impacts of aircraft operations. Above all, air transport represents one part of a much wider context: the considerable international efforts, since the 1980s, to develop ideas of sustainable development and to align economic, social and environmental activities with broader sustainable development objectives.

In this chapter, those issues are reviewed briefly and the main challenges faced by policymakers (and others) in attempting to reconcile air transport, economic growth, social development, environmental protection and sustainable development are summarised. As stated in Chapter 1, the aims of this book are: (a) to provide an updated overview of the environmental issues associated with air transport, emphasising recent scientific and policy developments; (b) to derive implications for the operation, growth, development and management of the air transport industry; (c) to inform aviation environmental policy; and (d) to provide a fresh analysis of the relationships between air transport, environmental protection and sustainable development, highlighting some areas in which further research is required. The contents of this book represent an attempt to meet those aims through accounts of aviation emissions and growth (Chapter 2), of the impacts of air transport on climate and air quality (Chapters 3 and 4), of the effects of aircraft noise (Chapter 5) and of the complex relationship between air transport and sustainable development (Chapter 6). Overall, a concerted attempt has been made to provide a constructive, policy-relevant synthesis of a wide range of perspectives rather than advocating one particular viewpoint. The main argument of this book, however, is that aviation environmental issues are inherently political (and should be acknowledged as such) – and that those issues should be interpreted in the context of a broader, nuanced and sufficiently sophisticated understanding of the principles and requirements of sustainable development. 'Sustainable development' is an ambiguous and highly-contested term that can mean 'practically nothing to people, or practically everything' (Adams 2009; Baker 2006; Dresner 2008; Elliott 2006; Jacobs 1991; O'Riordan 1988; Porritt 2005, 22; SDC 2001b). Yet the popularity of the concept of sustainable development has led some organisations and individuals to adopt the rhetoric of 'sustainability' with little clear idea of its provenance or implications. Hence terms such as 'sustainable aviation', 'sustainable mobility', 'sustainable society' and 'sustainable growth' have become commonplace in the literature of aviation environmental issues, although those terms may be extremely problematic. Understanding sustainable development is crucial to the task of understanding aviation environmental issues; ultimately, the effective management of aviation environmental impacts requires that the growth of air transport is managed in a way that is consistent with a sufficiently strong version of sustainable development.

A Variety of Challenges

The multitude of issues associated with the subject of air transport and the environment gives rise to a variety of different (although interrelated) challenges for policymakers, regulators and representatives of the air transport industry. Those challenges have been acknowledged in many places throughout this book; the main challenges are summarised briefly below. In many ways, aviation growth is the fundamental challenge underlying all of the others. The broad popularity of air transport and its importance in driving economic growth (and social development) mean that it is extremely difficult for decision-makers and policymakers to restrain the growth of aviation. Yet, ultimately, the management of all of the main environmental impacts of aviation requires precisely that: the restraint of aviation growth. Inevitably, therefore, debates about the future of air transport must focus on the level of restraint to be applied to aviation growth, the basis for choosing that level, the manner in which restraint is to be achieved and the implications of restraining aviation growth. Simultaneously, other challenges relating to air transport and the environment must be tackled. The challenge of climate change requires that dramatic cuts are made in greenhouse gas emissions, especially carbon dioxide (CO_2). The air transport industry also faces other climate-related issues, especially the need to allocate emissions from international aviation on an equitable basis between nations, and the need to reduce the non-CO_2 effects of aviation on climate. The challenge of meeting air quality objectives involves reducing emissions of key air pollutants – especially nitrogen oxides (NO_x) and particles – in the vicinity of airports: a task that, at least for NO_x, is very difficult to achieve whilst also reducing CO_2 emissions. Managing the impact of air transport on air quality also means tackling the problem of air pollution from aviation-related surface transport, a problem that is also exacerbated by aviation growth. The challenge of aircraft noise is a complex issue that requires a variety of approaches: the reduction of aircraft noise at source; the reduction of overall noise exposure in the vicinity of airports; the use of more effective methods of communicating noise exposure to those likely to be affected by its impacts; and the increased participation of those affected by aircraft noise in decisions about the operation and development of airports. Finally, the challenge of sustainable development is an overarching concern that requires the economic, social and environmental benefits and costs of air transport to be assessed in an integrated manner. Sustainable development also requires that the management of aviation environmental impacts does not blight economic development where it is most needed. More radically, air transport could contribute to sustainable development through its unique potential to assist in poverty reduction.

The Challenge of Aviation Growth

As explained in Chapter 2, the growth of air transport is a crucial dimension of the overall environmental impact of aviation. Since 1960, the aviation industry has experienced consistently high growth rates – of around 5 per cent per year – which have exceeded the rate of growth of the world economy (Bailey 2007; Humphreys 2003; IPCC 1999; Lee 2004; Lee et al. 2009). Aviation growth has been driven by a variety of factors including economic growth, social change, tourism growth and processes of globalisation (Bieger and Wittmer 2006; Daley et al. 2008; Debbage 1994; Goetz and Graham 2004; Gössling and Peeters 2007; Graham et al. 2008; Hettne 2008; Janelle and Beuthe 1997; May 2002; Mayor and Tol 2010; Pels 2008; Young 1997). A range of commercial market forecasts and authoritative scenarios of aviation growth have been produced; those forecasts and scenarios indicate that air transport is likely to continue to grow strongly over the decadal timescale – as will most of its emissions. Indeed, many forecasts and scenarios suggest that strong, sustained growth in demand for air transport is likely to continue to beyond 2030; by 2050, air passenger traffic is expected to have increased five-fold from 1995 levels (Bieger et al. 2007; Bows et al. 2005; 2006; DfT 2003b; IPCC 1999; Lee et al. 2009). Therefore, the growth of air transport is transforming the industry from being a relatively minor polluter into being a highly significant source of emissions, especially of CO_2. Even using a conservative estimate of the growth of aviation-related emissions, the quantity of CO_2 emitted by air transport is expected to quadruple over the period 1990–2050, and some forecasts suggest that CO_2 emissions from air transport could increase ten-fold over that period (Cairns and Newson 2006). Depending on the scenario used, the air transport industry could consume most – or even all – of national carbon budget allowances under the Kyoto Protocol or its successor agreement (Bows et al. 2009). The environmental impacts of air transport therefore require interpretation in the context of the rapid growth of aviation and most of its emissions. The influence of aviation growth is compounded by the long lead-in and in-service times associated with aviation infrastructure – with the implication that the task of reducing the projected environmental impacts of air transport is already an urgent one.

The Challenge of Climate Change

Chapter 3 has provided an account of the effects of air transport on climate. Climate change is now acknowledged to be one of the most prominent global environmental issues and coordinated efforts are required at all levels to adapt to, and to mitigate, its impacts (Adams 2009; Houghton 2009; IPCC 2007; Stern 2007). The air transport industry currently makes a small but nonetheless significant contribution to climate change, mainly due to the various effects of aircraft emissions; furthermore, that contribution is expected to increase with the rapid growth of aviation (Horton 2006; Houghton 2009; Lee 2004; Mayor and Tol 2010). As a result, by 2050, aircraft emissions are likely to constitute a substantial share of total greenhouse gas emissions worldwide, and aviation-related emissions could negate most or even all of the emissions reductions achieved by other sectors of national economies (Bows et al. 2009). For that reason, Houghton (2009, 346) has stated that 'controlling the growing influence of aviation on the climate is probably the largest challenge to be solved in the overall mitigation of climate change.' Yet the task of reducing the impact of air transport on climate represents a major challenge, not least because the industry is technologically mature, is characterised by long lead-in times and in-service times, is international in its scope and plays a vital role in promoting economic growth (Gössling and Upham 2009; Lee 2004; Peeters et al. 2009). Furthermore, aircraft affect climate in a variety of interrelated ways and, whilst some of those effects are well understood, others are still associated with substantial scientific uncertainties. In the short to medium term, the management of the impact of air transport on climate requires the formulation and implementation of effective policy, due to the high abatement costs of the sector and the limited potential for radical technological or operational solutions to be found, at least over the decadal timescale. To date, policy approaches have focused on the inclusion of aviation within emissions trading schemes such as the EU Emissions Trading Scheme (ETS); in addition, various voluntary commitments have been made within the industry, such as agreements to use carbon offsetting schemes or to achieve 'carbon neutrality'. However, many controversial issues are associated with such approaches – including the issue of how to guarantee CO_2 emission reductions and the question of how to deal with the other, non-CO_2 effects of aircraft on climate. Other policy instruments could also be used to reduce the impact of air transport on climate: the imposition of more stringent emissions limits; the removal of existing privileges and subsidies of the industry; the wider use of emissions charges, fuel taxes and other levies; and, ultimately, the use of severe demand restraint measures. Given their varying strengths and political acceptability, no single policy instrument appears to be ideal and an adequate policy framework will probably require the use of a combination of regulatory, market-based and voluntary approaches. The management of the impact of air transport on climate inevitably requires trade-offs to be made between different climate – and other environmental – effects. In the long term, the development of the air transport industry depends critically upon finding radical technological solutions to reduce the impact of air transport on climate.

The Challenge of Air Quality

In Chapter 4, the impact of air transport on air quality in the vicinity of airports was considered. Many aviation-related emissions degrade air quality; the substances of greatest concern for air quality management are NO_x and particles, since those emissions may cause air quality standards to be exceeded, with potentially serious implications for human health. The effects of another pollutant – hydrocarbons (HCs) – are currently little-known but are the subject of increasing concern and scientific investigation. Overall, most aviation-related emissions are increasing alongside the rapid growth of air transport, because that growth is outpacing the rate of technological and operational improvements in aircraft environmental performance. Aviation growth also compounds air pollution due to the associated increase in demand for surface transport. Consequently, air quality standards may constrain airport growth (and aviation growth more generally), especially as those standards become more stringent with time in response to decreasing public tolerance of environmental impacts; several European airports are already subject to stringent air quality constraints. With the introduction in the EU, in 2010, of mandatory air quality limits for nitrogen dioxide (NO_2) together with the ambitious long-term goals set by ICAO and ACARE to reduce NO_x emissions, the need to manage aviation impacts on air quality is becoming increasingly pressing, at least at the local level (DfT 2006; EEA 2006; Peace et al. 2006; Unal et al. 2005). Technological approaches to the reduction of aviation emissions focus on achieving specific NO_x reductions through improved engine design, on the use of alternative fuels and on operational measures to improve fuel efficiency. Yet those measures are not sufficient to offset the effects of rapid aviation growth; thus an effective policy response to the impact of air transport on air quality is required. Besides the ICAO aircraft engine certification standards, policy instruments have been little used to manage the impact of air transport on air quality and aviation emissions are not directly regulated at most airports. Policy approaches could focus on encouraging greater use of the most fuel-efficient (and least-polluting) aircraft, especially at major airports where compliance with air quality objectives is already marginal. Overall, a combination of policy approaches could be used: the promotion of voluntary agreements within the industry to reduce emissions; the imposition of more stringent emissions standards, especially for NO_x and particles; the removal of existing privileges and subsidies of the industry; the wider use of emissions charges, fuel taxes and other levies; the development of emissions trading schemes for key air pollutants; and, ultimately, the use of demand restraint measures and the restriction of airport infrastructure development. Thus the development of the air transport industry depends critically on finding radical technological solutions to reduce the NO_x and particle emissions associated with flight. A confounding factor, however, is that the 'technological fix' sought by the air transport industry – if current growth trends are to continue – must somehow reconcile reductions in NO_x and particle emissions with concomitant reductions in CO_2 emissions.

The Challenge of Aircraft Noise

Aircraft noise – discussed in Chapter 5 – is one of the most objectionable local environmental impacts of aviation (DfT 2002; 2003b; 2007; FAA 1985; Hume and Watson 2003; Hume et al. 2003; Kryter 1967; May and Hill 2006; Nero and Black 2000; Pattarini 1967; Skogö 2001; Thomas and Lever 2003). Aircraft noise causes annoyance to people by interrupting their communication and leisure activities, by disrupting activities requiring concentration and by discouraging people from using outdoor spaces. Aircraft noise also disturbs many people's sleep. Furthermore, aircraft noise may exacerbate conditions of stress, anxiety and ill-health for many people, especially those within more vulnerable social groups. As a result of those wide-ranging effects, the impact of aircraft noise can severely reduce people's wellbeing and quality of life (Hume and Watson 2003). Whilst advances have been made in airframe and engine design since the 1960s, with the result that individual aircraft movements are quieter, those advances are now being offset by the dramatic growth of air traffic (Hume and Watson 2003; Thomas and Lever 2003). Consequently, aircraft noise exposure around many airports is increasing and the annoyance caused to people is now due as much to the frequency with which people are over-flown as to the sound level of individual aircraft movements (Smith 1992; Thomas and Level 2003). Furthermore, as average affluence levels rise in many countries, individuals become increasingly sensitive to environmental nuisance and concerns about aircraft noise may generate fierce community opposition to airport operations and infrastructure development proposals. Thus public concerns about aircraft noise represent a major constraint to the growth of the air transport industry (EUROCONTROL 2004; 2008; Hume and Watson 2003; ICAO 1993; Moss et al. 1997). Managing the impact of aircraft noise requires achieving ongoing, substantial reductions in aircraft noise at source through innovations in airframe and engine technologies and through the use of revised operational procedures. Yet such technological and operational improvements will not offset the growing noise nuisance resulting from rapid aviation growth. Some regions, such as the EU, have demanded greater flexibility to exert stricter control over aircraft noise through the use of noise-related regulations, although that suggestion raises important issues of equity: the imposition of stringent noise-related restrictions could disproportionately affect the airlines of developing countries – those with the least capacity to undertake fleet renewal. The issue of aircraft noise is thus a complex, subjective topic that raises many questions about perception, representation, tolerance, affluence, power and justice (Maris et al. 2007; May and Hill 2006; Thomas and Lever 2003). Again, inevitably, trade-offs must be made between the reduction of aircraft noise and the management of other environmental issues – especially climate change.

The Challenge of Sustainable Development

Analysis of the environmental impacts of air transport raises a critical question: should aviation growth be constrained in order to keep aviation environmental impacts within acceptable limits, or should aviation growth be permitted – and its intensifying environmental impacts justified – on other (economic and social) grounds? As explained in Chapter 6, that question lies at the heart of debates about air transport and sustainable development. Ultimately, that question is about values, attitudes and beliefs – and also about the ways in which conflicting views are acknowledged and (ideally) reconciled. Environmental issues now tend to be interpreted within the context of broader debates about sustainable development, a concept that has become the central organising principle in discussions about how to manage environmental impacts whilst ensuring that the development of human economies and societies occurs in an equitable way (Adams 2009). In turn, the concept of sustainable development incorporates other, core principles, particularly those of intergenerational and intragenerational equity, poverty reduction and participatory governance (Adams 2009; Baker 2006; Dresner 2008; Elliott 2006; Jacobs

1995; Redclift 1984). Whilst sustainable development is acknowledged to be a complex, ambiguous term with a wide range of meanings, the original formulation by the World Commission on Environment and Development (WCED) retains its authoritative status (Baker 2006; WCED 1987). As concerns about global environmental change – especially climate change – become acute, the imperative for air transport to operate on a basis that promotes sustainable development is becoming increasingly clear. The task of aligning the operation of the air transport industry with the principles and requirements of sustainable development represents a formidable challenge – one that necessitates the effective management of aviation growth. Nonetheless, making air transport consistent with sustainable development is the only way in which a compromise between aviation growth (and economic growth more generally) and environmental protection may be found that is acceptable to the majority of people. One important consideration is that the management of aviation environmental impacts – whilst essential – should not blight economic development where it is most needed. Yet, more creative possibilities for the relationship between air transport and sustainable development can also be envisioned and explored. In particular, the air transport industry has unique potential to contribute to poverty reduction worldwide – and thereby to promote sustainable development. The potential of the air transport industry to contribute to a dramatic reduction in poverty worldwide is one of the most exciting possibilities to emerge from recent debates about the relationships between air transport, economic growth, environmental protection and sustainable development.

'Dissenting Imaginations'

For too long, debates about air transport and the environment have rehearsed familiar arguments between those who promote aviation growth and those who advocate environmental protection. Arguably, those debates represent little change from the bitter conflicts of the 1970s and 1980s between developmentalists and environmentalists – the context in which the authoritative, Brundtland definition of sustainable development emerged (Baker 2006; WCED 1987). Despite the fact that remarkable advances have been made, recently, in understanding the scientific basis of aviation environmental impacts, relatively little progress has been made in clarifying the nature of the relationship between air transport and sustainable development. Indeed, the concepts of 'sustainability' and sustainable development have been used by both sides – developmentalists and environmentalists – to justify their entrenched positions. It is not surprising that debates about air transport and the environment are so contentious: they relate to widespread attitudes about choice, freedom and individuality; they expose and critique dominant patterns of consumption; and they call into question widely-held notions of wellbeing and quality of life. Thus the controversy surrounding aviation environmental issues underlines the fact that those issues are essentially political – and that the task of managing those issues is an inherently political task. Yet, if debates about air transport and the environment are to progress beyond the now-familiar arguments, it is important to dispense with simple binary oppositions: development versus environment; growth versus stagnation; culture versus nature. In particular, a much more realistic view of aviation growth is required. Aviation growth cannot continue indefinitely and it should not be allowed to proceed unconstrained; yet it is possible to conceive of a much more sophisticated approach towards aviation growth, with growth managed at an appropriate level and with the benefits of air transport being carefully directed to where they are most needed. Therefore, the management of aviation growth should be placed within the context of broader, creative visions of the future of air transport. For this, we need 'dissenting imaginations' to envision and explore the most promising ideas of what the air transport industry could become (Escobar 2004). Such a future air transport system would ideally be an integral part of healthy societies and economies, contributing widely to human wellbeing and quality of life, rather than being a relentlessly-expanding activity whose benefits are enjoyed exclusively by an affluent minority of the global population. Poverty reduction could – and should – be part of that vision.

References

Abeyratne, R.I.R. 1999, 'Management of the environmental impact of tourism and air transport on small island developing states', *Journal of Air Transport Management*, 5: 31–37.

Abeyratne, R.I.R. 2003, 'Air transport and sustainable development', *Environmental Policy and Law*, 33(3–4): 138–43.

Abeyratne, R.I.R. 2004, *Aviation in Crisis*, Aldershot, Ashgate Publishing Limited.

ACARE (Advisory Council for Aeronautics Research in Europe) 2004, *Strategic Research Agenda*, vol. 1, October 2004, Brussels, ACARE.

Adams, W.M. 2009, *Green Development: Environment and Sustainability in a Developing World*, 3rd edition, London, Routledge.

Adey, P., Budd, L. and Hubbard, P. 2007, 'Flying lessons: exploring the social and cultural geographies of global air travel', *Progress in Human Geography*, 31(6): 773–91.

AEA Technology Environment 2004, *An Evaluation of the Air Quality Strategy: Final Report to DEFRA*, December 2004, Didcot, AEA Technology plc.

Air Travel – Greener By Design 2005, *Annual Report 2005–2006*, London, Royal Aeronautical Society.

Air Travel – Greener By Design 2006, *Annual Report 2005–2006*, London, Royal Aeronautical Society.

Air Travel – Greener By Design 2007, *Annual Report 2006–2007*, London, Royal Aeronautical Society.

Air Travel – Greener By Design 2008, *Annual Report 2007–2008*, London, Royal Aeronautical Society.

Airbus 2006, *Global Market Forecast: The Future of Flying 2006–2025*, Blagnac Cedex, Airbus.

Airbus 2007, *Flying By Nature: Global Market Forecast 2007–2026*, Blagnac Cedex, Airbus.

Åkerman, J. 2005, 'Sustainable air transport – on track in 2050', *Transportation Research Part D*, 10, 111–26.

Allen, J.E. 1999, 'Global energy issues affecting aeronautics: a reasoned conjecture', *Progress in Aerospace Sciences*, 35: 413–53.

Amanatidis, G. and Friedl, R. 2004, 'Science related to atmospheric effects of aircraft emissions continues to mature', *ICAO Journal*, 59 (5): 14–15, 26.

Anderson, B.E., Chen, G. and Blake, D.R. 2006, 'Hydrocarbon emissions from a modern commercial airliner', *Atmospheric Environment*, 40: 3601–12.

Anderson, K.L., Mander, S.L., Bows, A., Shackley, S., Agnolucci, P. and Ekins, P. 2008, 'The Tyndall decarbonisation scenarios – Part II: scenarios for a 60% CO_2 reduction in the UK', *Energy Policy*, 36(10): 3764–73.

Anderson, K., Shackley, S., Mander, S. and Bows, A. 2005, *Decarbonising the UK: Energy for a Climate Conscious Future*, Manchester, Tyndall Centre for Climate Change Research.

Andrieu, M. 1993, 'What horizons for air transport?', *OECD Observer*, 180: 4–7.

Arafa, M.H., Osman, T.A. and Abdel-Latif, I.A. 2007, 'Noise assessment and mitigation schemes for Harghada airport', *Applied Acoustics*, 68(11–12): 1373–85.

Archer, D. 2005, 'Fate of fossil fuel CO_2 in geologic time', *Journal of Geophysical Research*, 110: C09S05, doi:10.1029/2004JC002625.

Arrhenius, S. 1896, 'On the influence of carbonic acid in the air upon the temperature of the ground', *Philosophical Magazine*, 4: 237–76.

ATAG (Air Transport Action Group) 2002, *Industry as a Partner for Sustainable Development*, Geneva, ATAG.

ATAG 2009, '*Air transport – a global approach to sustainability*' [online], available at http://www.airport-int.com/categories/environment/air-transport-a-global-approach-to-sustainability.asp, accessed 14 May 2009.

Auerbach, S. and Koch, B. 2007, 'Cooperative approaches to managing air traffic efficiently – the airline perspective', *Journal of Air Transport Management*, 13: 37–44.

Babisch, W. 1998, 'Epidemiological studies of the cardiovascular effects of occupational noise: a critical appraisal', *Noise and Health*, 1: 24–39.

Babisch, W. 2000, 'Traffic noise and cardiovascular disease: epidemiological review and synthesis', *Noise and Health*, 8: 9–32.

Bailey, J.W. 2007, 'An assessment of UK Government aviation policies and their implications', *Journal of Airport Management*, 1(3): 249–61.

Baker, S. 2006, *Sustainable Development*, Routledge Introductions to Environment Series, London, Routledge.

Bayon, R., Hawn, A. and Hamilton, K. 2007, *Voluntary Carbon Markets: An International Business Guide to What They Are and How They Work*, London, Earthscan.

Becken, S. 2007, 'Tourists' perception of international air travel's impact on the global climate and potential climate change policies', *Journal of Sustainable Tourism*, 15(4): 351–68.

Ben-Yosef, E. 2005, *The Evolution of the US Airline Industry: Theory, Strategy and Policy*, Studies in Industrial Organization, New York, Springer.

Berckmans, D., Janssens, K., Van der Auweraer, Sas, P. and Desmet, W. 2008, 'Model-based synthesis of aircraft noise to quantify human perception of sound quality and annoyance', *Journal of Sound and Vibration*, 311(3–5): 1175–95.

Berry, B.F. and Jiggins, M. 1999, *An Inventory of UK Research on Noise and Health, 1994–1999*, National Physics Laboratory, NPL Report CMAM 40, Teddington, NPL.

Bieger, T. and Wittmer, A. 2006, 'Air transport and tourism – perspectives and challenges for destinations, airlines and governments', *Journal of Air Transport Management*, 12: 40–46.

Bieger, T., Wittmer, A. and Laesser, C. 2007, 'What is driving the continued growth in demand for air travel? Customer value of air transport', *Journal of Air Transport Management*, 13: 31–6.

Bishop, S. and Grayling, T. 2003, *The Sky's the Limit: Policies for Sustainable Aviation*, London, Institute for Public Policy Research.

Boeing 2006, *Current Market Outlook 2006*, Seattle, Boeing.

Boeing 2008, *Current Market Outlook 2008–2027*, Seattle, Boeing.

Boon, B.H. and Wit, R.C.N. 2005, *The Contribution of Aviation to the Economy: Assessment of Arguments Put Forward*, Delft, CE.

Borsky, P.N. 1979, 'Sociopsychological factors affecting the human response to noise exposure', *Otolaryngologic Clinics of North America*, 12: 521–535.

Boucher, O. 1999, 'Air traffic may increase cirrus cloudiness', *Nature*, 397: 30–31.

Bows, A., Anderson, K. and Footitt, A. 2009, Aviation in a low-carbon EU, in Gössling, S. and Upham, P. (eds) *Climate Change and Aviation: Issues, Challenges and Solutions*, London, Earthscan: 89–109.

Bows, A., Anderson, K. and Upham, P. 2006, *Contraction and Convergence: UK Carbon Emissions and the Implications for UK Air Traffic*, Manchester, Tyndall Centre for Climate Change Research.

Bows, A., Anderson, K. and Upham, P. 2009, *Aviation and Climate Change: Lessons for European Policy*, Routledge Studies in Physical Geography and Environment, London, Routledge.

Bows, A., Upham, P. and Anderson, K. 2005, *Growth Scenarios for EU and UK Aviation: Contradictions with Climate Policy*, Summary of research by the Tyndall Centre for Climate Change Research for Friends of the Earth Trust Ltd, 16 April 2005, Manchester, Tyndall Centre for Climate Change Research.

Brasseur, G.P., Cox, R.A., Hauglustaine, D., Isaksen, I., Lelieveld, J., Lister, D.H., Sausen, R., Schumann, U., Wahner, A. and Wiesen, P. 1998, 'European scientific assessment of the atmospheric effects of aircraft emissions', *Atmospheric Environment*, 32: 2329–2418.

Broderick, J. 2009, Voluntary carbon offsetting for air travel, in Gössling, S. and Upham, P. (eds) *Climate Change and Aviation: Issues, Challenges and Solutions*, London, Earthscan: 329–46.

Bronzaft, A.L., Ahern, K.D., McGinn, R. and Savino, B. 1998, 'Aircraft noise: a potential health hazard', *Environment and Behaviour*, 30(1): 101–13.

Brouwer, R., Brander, L. and van Beukering, P. 2007, '"A convenient truth": air travel passengers' willingness to pay to offset their CO_2 emissions', Amsterdam: Institute for Environmental Studies, Vrije Universiteit.

Brueckner, J.K. and Girvin, R. 2008, 'Airport noise regulation, airline service quality, and social welfare', *Transportation Research Part B: Methodological*, 42(1): 19–37.

Budd, L. and Graham, B. 2009, 'Unintended trajectories: liberalization and the geographies of private business flight', *Journal of Transport Geography*, 17(4): 285–92.

Button, K. 2004, *Wings Across Europe: Towards an Efficient European Air Transport System*, Aldershot, Ashgate Publishing Limited.

CAA (Civil Aviation Authority) 2006, *ICAO Aircraft Engine Exhaust Emissions Data Bank* [online], available at http://www.caa.co.uk/docs /702/050601_PC_ICAO_Engine_Emissions_Databank-Issue_14.xls, accessed 7 August 2006.

CAA 2008, *CAA Sustainable Development and Aviation Environmental Policy*, London, CAA.

Cairns, S. and Newson, C. 2006, *Predict and Decide: Aviation, Climate Change and UK Policy*, Oxford, Environmental Change Institute.

Campling, L. and Rosalie, M. 2006, 'Sustaining social development in a small island developing state? The case of Seychelles', *Sustainable Development*, 14: 115–25.

Carlsson, F. and Hammar, H. 2002, 'Incentive based regulation of CO_2 emissions from international aviation', *Journal of Air Transport Management*, vol. 8: 365–72.

Carter, N.L. 1996, 'Transportation noise, sleep, and possible after-effects', *Environment International*, 22(1): 105–16.

Carter, N.L., Hunyor, S.N., Crawford, G., Kelly, D. and Smith, A.J.M. 1994a, 'A study of arousals, cardiac arrhythmias and urinary catecholamines', *Sleep*, 17: 927–42.

Carter, N.L., Ingham, P., Tran, K. and Hunyor, S.N. 1994b, 'A field study of the effects of traffic noise on heart rate and cardiac arrhythmias during sleep', *Journal of Sound and Vibration*, 169(2): 211–27.

Caves, R.E. 1994a, 'Aviation and society – redrawing the balance (I)', *Transportation Planning and Technology*, 18: 3–19.

Caves, R.E. 1994b, 'Aviation and society – redrawing the balance (II)', *Transportation Planning and Technology*, 18: 21–36.

Caves, R.E. 2003, The social and economic benefits of aviation, in Upham, P., Maughan, J., Raper, D.W. and Thomas, C.S. (eds), *Towards Sustainable Aviation*, London, Earthscan, 36–47.

CEC (Commission of the European Communities) 1999, *Air Transport and the Environment: Toward Meeting the Challenges of Sustainable Development*, Communication from the Commission to the Council, the European Parliament, the Economic and Social Committee and the Committee of the Regions, Brussels, CEC.

CEC 2006, *Proposal for a Directive of the European Parliament and of the Council amending Directive 2003/87/EC so as to include aviation activities in the scheme for greenhouse gas emission allowance trading within the Community* [online], http://europa.eu/scadplus/leg/en/lvb/l28012.htm, accessed 12 June 2008.

Chaplin, V. 1996, 'Aggressive and proactive campaign has resulted in improvement in environmental performance', *ICAO Journal*, 51(2): 17–18.

Clarke, J.-P., 2003, 'The role of advanced air traffic management in reducing the impact of aircraft noise and enabling aviation growth', *Journal of Air Transport Management*, 9: 161–5.

Collins, P., Dowling, A. and Greitzer, E. 2006, 'Academia exploring innovative approaches to achieving "silent" flight', *ICAO Journal*, 61(1): 24–5, 37.

Connor, T.L. 1996, 'Proposed take-off noise abatement procedures demonstrate potential to mitigate problem', *ICAO Journal*, 51(2): 19–20.

Crayston, J. 1992, 'ICAO group identifies environmental problems associated with civil aviation', *ICAO Journal*, 47(8): 4.

Crayston, J. and Hupe, J. 1999, 'ICAO facing complex and evolving challenges in the environmental field', *ICAO Journal*, 54(7): 5–8.

Crutzen, P.J. 1971, 'Ozone production rate in an oxygen-hydrogen-nitrogen atmosphere', *Journal of Geophysical Research*, 76: 7311–27.

CSP (The Chartered Society of Physiotherapy) 2006, *Dangerous levels of toxic gas detected at most major airports* [online], available at http://www.csp.org.uk, accessed 30 March 2006.

Curran, R. 2006, *Correction to Engine Emissions Data Resulting From Engine Deterioration*, QinetiQ/05/01726, August 2006, Farnborough, QinetiQ.

Daley, B. 2008, 'Is air transport an effective tool for sustainable development?', *Sustainable Development*, 17(4): 210–19.

Daley, B., Dimitriou, D. and Thomas, C. 2008, The environmental sustainability of aviation and tourism, in Graham, A., Papatheodorou, A. and Forsyth, P. (eds), *Aviation and Tourism: Implications for Leisure Travel*, Aldershot, Ashgate Publishing Limited, 239–53.

Daley, B. and Preston, H. 2009, Aviation and climate change: assessment of policy options, in Gössling, S. and Upham, P. (eds) *Climate Change and Aviation: Issues, Challenges and Solutions*, London, Earthscan: 347–72.

Debbage, K.G. 1994, 'The international airline industry: globalization, regulation and strategic alliances', *Journal of Transport Geography*, 2(3): 190–203.

DEFRA (Department for Environment, Food and Rural Affairs) 2007a, *'Climate change: carbon offsetting – Code of Best Practice'* [online], http://www.defra.gov.uk/environment/climatechange/uk/carbonoffset/codeofpractice.htm, accessed 12 July 2007.

DEFRA 2007b, *The Air Quality Strategy for England, Scotland, Wales and Northern Ireland (Volume 1)*, July 2007, Cm 7169, Norwich, HMSO.

De Leon, R.R and Haigh, J.D. 2007, 'Infrared properties of cirrus clouds in climate models', *The Quarterly Journal of the Royal Meteorological Society*, 133(623): 273–82.

Dennis, N. 2009, Airline trends in Europe: network consolidation and the mainstreaming of low-cost strategies, in Gössling, S. and Upham, P. (eds) *Climate Change and Aviation: Issues, Challenges and Solutions*, London, Earthscan: 151–76.

Depitre, A. 2001, 'Re-certification of aircraft to new noise standards remains an important issue', *ICAO Journal*, 56(4): 14–16, 32.

DETR (Department of Environment, Transport and the Regions) 1999, *Report on the Review of the National Air Quality Strategy: Proposals to Amend the Strategy*, London, DETR.

DETR 2000, *The Air Quality Strategy for England, Scotland and Wales and Northern Ireland: Working Together for Clean Air*, Cm 4548, Norwich, The Stationary Office.

DfID (Department for International Development) 2007, *Millennium Development Goals* [online], available at http://www.dfid.gov.uk/mdg/, accessed 12 July 2007.

DfT (Department for Transport) 1992, *Report of a Field Study of Aircraft Noise and Sleep Disturbance*, A Study Commissioned by the Department of Transport from the Department of Safety, Environment and Engineering, Civil Aviation Authority, December 1992, Ollerhead, J.B., Jones, C.J., Cadoux, R.E., Woodley, A., Atkinson, B.J., Horne, J.A., Pankhurst, F., Reyner, L., Hume, K.I., Van, F., Watson, A., Diamond, I.D., Egger, P., Holmes, D. and McKean, J., London, DfT.

DfT 2002, *Regional Air Services Co-ordination Study, RASCO Final Report*, July 2002, London, DfT.

DfT 2003a, *Aviation, Core Cities and Regional Economic Development*, London, DfT.

DfT 2003b, *Night Flying Restrictions at Heathrow, Gatwick and Stansted*, London: DfT.

DfT 2003c, *The Future of Air Transport*, Presented to Parliament by the Secretary of State for Transport by Command of Her Majesty, December 2003, Cm 6046, London, DfT.

DfT 2004, *Aviation and Global Warming*, London, DfT.

DfT 2006, *Project for the Sustainable Development of Heathrow – Report of the Air Quality Technical Panels* [online], available at http://www.dft.gov.uk/pgr/aviation/environmentalissues/heathrowsustain/, accessed 21 May 2009.

DfT 2007, *ANASE: Attitudes to Noise from Aviation Sources in England*, Final Report for Department of Transport, MVA Consultancy in Association with John Bates Services, Ian Flindell and RPS, October 2007, London, DfT.

DfT 2008, *Aviation Emissions Cost Assessment 2008*, London, DfT.

Dobbie, L. 1999, 'Airlines see direct link between improved environmental performance, sustainable growth', *ICAO Journal*, 54(7): 15–17, 29.

Dobbie, L. and Eran-Tasker, M. 2001, 'Measures to minimise fuel consumption appear to be of greatest importance to airlines', *ICAO Journal*, 56(4): 24–5, 31–2.

Dobruszkes, F. 2006, 'An analysis of low-cost airlines and their networks', *Journal of Transport Geography*, 14: 249–64.

DOTARS (Department of Transport and Regional Services) 2000, *Expanding Ways to Describe and Assess Aircraft Noise*, Canberra, DOTARS.

Drake, F. and Purvis, M. 2001, 'The effect of supersonic transports on the global environment', *Science, Technology, and Human Values* 26(4): 501.

Drake, J.W. 1974, 'Social, political and economic constraints on airline fuel optimization', *Transportation Research*, 8: 443–9.

Dresner, S. 2008, *The Principles of Sustainability*, 2nd edition, London, Earthscan.

DTI (Department of Trade and Industry) 1996, 'Experts consider operational measures as means to reduce emissions and their environmental impact', *ICAO Journal*, 51(2): 9–10.

Dubois, G. and Ceron, J.-P. 2006, 'Tourism and climate change: proposals for a research agenda', *Journal of Sustainable Tourism* 14(4), 399–415.

EC 2002, *Directive 2002/30/EC of the European Parliament and of the Council of 26 March 2002 on the Establishment of Rules and Procedures with regard to the Introduction of Noise-Related Operating Restrictions at Community Airports*, 28 March 2002, Brussels, EC.

EC 2005, *Directive on Ambient Air Quality and Cleaner Air for Europe*, 2008/50/EC, Brussels, EC.

EC 2006, Climate change: Commission proposes bringing air transport into EU Emissions Trading Scheme, Press Release, IP/06/1862, 20 December 2006, Brussels, EC.

EC 2008, *Directive 2008/50/EC of the European Parliament and of the Council of 21 May 2008 on Ambient Air Quality and Cleaner Air for Europe*, Brussels, EC.

Eckersley, R. 1992, *Environmentalism and Political Theory: Towards an Ecocentric Approach*, London, UCL Press.

Edwards-Jones, G., Milà i Canals, L., Hounsome, N., Truninger, M., Koerber, G., Hounsome, B., Cross, P., York, E.H., Hospido, A., Plassmann, K., Harris, I.M., Edwards, R.T., Day, G.A.S., Tomos, A.D., Cowell, S.J. and Jones, D.L. 2008, 'Testing the assertion that 'local food is best': the challenges of an evidence-based approach', *Trends in Food Science and Technology*, 19(5): 265–74.

EEA (European Environment Agency) 2006, *Transport and Environment: Facing a Dilemma – TERM 2005*, EEA Report No 3/2006, Copenhagen, EEA.

EEA 2008, CSI 004 – *Exceedance of air quality limit values in urban areas (version 1) – Assessment published Apr 2008* [online], available at http://themes.eea.europa.eu/IMS/IMS/ISpecs/ISpecification200410 01123040/IAssessment1197036296175/view_content, accessed 23 May 2009.

Ekins, P. 2000, *Economic Growth and Environmental Sustainability: The Prospects for Green Growth*, London, Routledge.

Elliott, J.A. 2006, *Introduction to Sustainable Development*, 3rd edition, London, Routledge.

EPA (Environmental Protection Agency) 1995, *Compilation of Air Pollutant Emission Factors: Volume I: Stationary Point and Area Sources,* AP-42, 5th edition, January 1995, Research Triangle Park, NC, EPA.

EPA 1997, *Procedures for Preparing Emission Factor Documents*, EPA-454/R-95-015, revised edition, November 1997, Research Triangle Park, NC, EPA.

Escobar, A. 2004, 'Beyond the Third World: imperial globality, global coloniality and anti-globalisation social movements', *Third World Quarterly*, 25(1): 207–30.

EUROCONTROL 2004, *Challenges to Growth 2004 Report*, Brussels, EUROCONTROL.

EUROCONTROL 2007, *SESAR: Single European Sky ATM Research*, Brussels, EUROCONTROL.

EUROCONTROL 2008, *Challenges of Growth 2008: Summary Results*, Brussels, EUROCONTROL.

Evans, G.W. and Lepore, S.J. 1993, 'Nonauditory effects of noise on children: a critical review', *Children's Environments*, 10(1): 31–51.

Eyers, C.J., Norman, P., Middel, J., Plohr, M., Michot, S., Atkinson, K. and Christou, R.A. 2004, *AERO2k Global Aviation Emission Inventories for 2002 and 2025*, Farnborough, QinetiQ.

FAA (Federal Aviation Administration) 1985, *Aviation Noise Effects*, FAA-EE-85-2, March 1985, Washington, DC, FAA.

FAA 1997, *Air Quality Procedures for Civilian Airports and Air Force Bases*, Washington, DC, FAA.

FAA 2005, *Aviation Noise Effects*, ADA-154319 NTIS, Washington, DC, FAA.

Faber, J., Boon, N., Berk, M., den Elzen, M., Olivier, J. and Lee, D. 2007, *Climate Change Scientific Assessment and Policy Analysis: Aviation and Maritime Transport in a Post 2012 Climate Policy Regime*, Delft, CE.

Fawcett, A. 2000, 'The sustainability of airports and aviation: depicturing air travel, impacts and opportunities for sustainable change', *Transport Engineering in Australia*, 6: 33–9.

Fenley, C.A., Machado, W.V. and Fernandes, E. 2007, 'Air transport and sustainability: lessons from Amazonas', *Applied Geography*, 27: 63–77.

Ferrar, T.A. 1974, 'The allocation of airport capacity with emphasis on environmental quality', *Transportation Research*, 8: 163–9.

Fichter, C., Marquart, S., Sausen, R. and Lee, D.S. 2005, 'The impact of cruise altitude on contrails and related radiative forcing', *Meteorologische Zeitschrift*, 14: 563–72.

Fidell, S. 1999, 'Assessment of the effectiveness of aircraft noise regulation', *Noise and Health*, 3: 17–25.

Fidell, S., Pearsons, K. Tabachnick, B., Howe, R., Silvati, L. and Barber, L.D. 1995, 'Field study of noise-induced sleep disturbance', *Journal of the Acoustical Society of America*, 98(2): 1025–33.

Finley, M. 2008, 'Composites make for greener aircraft engines', *Reinforced Plastics*, 52(1): 24–5, 27.

Flindell, I. and Stallen, P.J. 1999, 'Non-acoustical factors in environmental noise', *Noise and Health*, 3: 11–16.

Forster, P.M. de F., Shine, K.P. and Stuber, N. 2006, 'It is premature to include non-CO_2 effects of aviation in emission trading schemes', *Atmospheric Environment*, 40: 1117–21.

Forsyth, P. 2007, 'The impacts of emerging aviation trends on airport infrastructure', *Journal of Air Transport Management*, 13: 45–52.

Forsyth, P. 2008, Tourism and aviation policy: exploring the links, in Graham, A., Papatheodorou, A. and Forsyth, P. (eds), *Aviation and Tourism: Implications for Leisure Travel*, (ed.) London, Earthscan, 73–82.

Franken, W. 2001, 'Experts propose more stringent standards for noise from large jets and propeller-driven aircraft', *ICAO Journal*, 56(4): 8–9, 31.

Freer, D.W. 1994, 'ICAO at 50 years: riding the flywheel of technology', *ICAO Journal*, 49:7, 19–32.

Friedl, R.R., Baughcum, S.L., Anderson, B., Hallett, J., Liou, K.-N., Rasch, P., Rind, D., Sassen, K., Singh, H., Williams, L. and Wuebbles, D. 1997, *Atmospheric Effects of Subsonic Aircraft: Interim Assessment of the Advanced Subsonic Assessment Program*, NASA Reference Publication 1400, Washington, DC, NASA.

Friends of the Earth, Greenpeace and WWF-UK. 2006, Joint statement on offsetting carbon emissions, London, Friends of the Earth, Greenpeace and WWF-UK.

Gander, S. and Helm, N. 1999, 'Emissions trading is an effective, proven policy tool for solving air pollution problems', *ICAO Journal*, 54(7): 12–14, 28–9.

Garvey, J.F. 2001, 'Complex noise issue calls for environmentally and economically responsible solution', *ICAO Journal*, 56(4): 20–21, 33–4.

Gauss, M., Isaksen, I.S.A., Wong, S. and Wang, W.-C. 2003, 'The impact of H_2O emissions from kerosene aircraft and cryoplanes on the atmosphere', *Journal of Geophysical Research*, 108, doi:10.1029/2002JD002623.

Gibbon, P. and Bolwig, S. 2007, *The Economic Impact of a Ban on Imports of Air Freighted Organic Products to the UK*, October 2007, Copenhagen, Danish Institute for International Studies.

Gierens, K. 2007, 'Are fuel additives a viable contrail mitigation option?' *Atmospheric Environment*, 41(21): 4548–52.

Gierens, K., Lim, L. and Eleftheratos, K. 2008, 'A review of various strategies for contrail avoidance', *The Open Atmospheric Science Journal*, 2: 1–7.

Gillen, D. and Lall, A. 2004, 'Competitive advantage of low-cost carriers: some implications for airports', *Journal of Air Transport Management*, 10: 41–50.

Gillen, D.W. and Levesque, T.J. 1994, 'A socio-economic assessment of complaints about airport noise', *Transportation Planning Technology*, 13: 45–5.

Girvin, R. 2008, 'Aircraft noise-abatement and mitigation strategies', *Journal of Air Transport Management*, 15(1): 14–22.

Givoni, M. and Rietveld, P. 2009, 'Airlines' choice of aircraft size – explanations and implications', *Transportation Research Part A: Policy and Practice*, 43(5): 500–10.

Goetz, A.R. and Graham, B. 2004, 'Air transport globalization, liberalization and sustainability: post-2001 policy dynamics in the United States and Europe', *Journal of Transport Geography*, 12: 265–76.

Goldstein A. 2001, 'Infrastructure development and regulatory reform in sub-Saharan Africa: the case of air transport', *World Economy*, 24: 221–3.

Gössling, S. 2002, 'Global environmental consequences of tourism', *Global Environmental Change*, 12(4): 283–302.

Gössling, S., Broderick, J., Upham, P., Ceron, J.-P., Dubois, G., Peeters, P. and Strasdas, W. 2007, 'Voluntary carbon offsetting schemes for aviation: efficiency, credibility and sustainable tourism', *Journal of Sustainable Tourism*, 15(3): 223–48.

Gössling, S., Ceron, J.-P., Dubois, G. and Hall, M.C. 2009, Hypermobile travellers, in Gössling, S. and Upham, P. (eds), *Climate Change and Aviation: Issues, Challenges and Solutions*, London, Earthscan: 131–50.

Gössling, S. and Peeters, P. 2007, '"It does not harm the environment!" An analysis of industry discourses on tourism, air travel and the environment', *Journal of Sustainable Tourism*, 15(4): 402–17.

Gössling, S., Peeters, P., Ceron, J.-P., Dubois, G., Patterson, T. and Richardson, R. 2005, 'The eco-efficiency of tourism', *Ecological Economics*, 54(4): 417–34.

Gössling, S. and Upham, P. 2009, Introduction: aviation and climate change in context, in Gössling, S. and Upham, P. (eds) *Climate Change and Aviation: Issues, Challenges and Solutions*, London, Earthscan: 1–23.

Gough, I. and McGregor, J.A. (eds) 2007, *Wellbeing in Developing Countries: From Theory to Research*, Cambridge, Cambridge University Press.

Graham, A. 2001, *Managing Airports: An International Perspective*, Oxford, Butterworth-Heinemann.

Graham, A. 2008, Trends and characteristics of leisure travel demand, in Graham, A., Papatheodorou, A. and Forsyth, P. (eds), *Aviation and Tourism: Implications for Leisure Travel*, London, Earthscan, 21–33.

Graham, A., Papatheodorou, A. and Forsyth, P. 2008, Introduction, in Graham, A., Papatheodorou, A. and Forsyth, P. (eds), *Aviation and Tourism: Implications for Leisure Travel*, London, Earthscan, 1–4.

Graham, A. and Raper, D.W. 2006a, 'Transport to ground of emissions in aircraft wakes, Part I: processes', *Atmospheric Environment*, 40: 5574–85.

Graham, A. and Raper, D.W. 2006b, 'Transport to ground of emissions in aircraft wakes, Part II: effect on NOx concentrations in airport approaches', *Atmospheric Environment*, 40: 5824–36.

Graham, B. 1995, *Geography and Air Transport*, Chichester, Wiley.

Graham, B. and Guyer, C. 1999, 'Environmental sustainability, airport capacity and European air transport liberalization: irreconcilable goals?', *Journal of Transport Geography*, 7: 165–80.

Graham, B. and Shaw, J. 2008, 'Low-cost airlines in Europe: reconciling liberalization and sustainability', *Geoforum*, 39(3): 1439–51.

Greaves, S. and Collins, A. 2007, 'Disaggregate spatio-temporal assessments of population exposure to aircraft noise', *Journal of Air Transport Management*, 13(6): 338–47.

Grewe, V., Dameris, M., Fichter, C. and Lee, D.S. 2002, 'Impact of aircraft NOx emissions: Part 2: effects of lowering the flight altitude', *Meteorologische Zeitschrift*, 11: 197–205.

Griefahn, B. and Muzet, A. 1978, 'Noise-induced sleep disturbances and their effects on health', *Journal of Sound and Vibration*, 59(1): 99–106.

Grubesic, T. and Zook, M. 2007, 'A ticket to ride: evolving landscapes of air travel accessibility in the United States', *Journal of Transport Geography*, 15: 417–30.

Hajer, M.A. 1996, Ecological modernization as cultural politics, in Lash, S., Szerzynski, B. and Wynne, B. (eds), *Risk, Environment and Modernity: Towards a New Ecology*, London, Sage Publications, 246–68.

Hanks, K.P. 2006, 'How should airports be regulated?', *Airport Management*, 1(1): 1–9.

Haralabidis, A.S., Dimakopoulou, K., Vigna-Taglianti, F., Giampaolo, M., Borgini, A., Dudley, M.-L., Pershagen, G., Bluhm, G., Houthuijs, D., Babisch, W., Velonakis, M., Katsouyanni, K. and Jarup, L. 2008, 'Acute effects of night-time noise exposure on blood pressure in populations living near airports', *European Heart Journal*, 29(5): 658–64.

Hartwick, E. and Peet, R. 2003, 'Neoliberalism and nature: the case of the WTO', *Annals of the American Academy of Political Economy*, 590: 188–211.

Hatfield, J., Job, R.F.S., Carter, N.L., Peploe, P., Taylor, R. and Morrell, S. 2001, 'The influence of psychological factors on self-reported physiological effects of noise', *Noise and Health*, 3(10): 1–13.

Health Canada 2001, *Noise from Civilian Aircraft in the Vicinity of Airports – Implications for Human Health*, Volume I: Noise, Stress and Cardiovascular Disease, Ottawa, Health Canada.

Health Council of the Netherlands 1999, *Public Health Impact of Large Airports*, Report of the Committee on the Health Impact of Large Airports, 1999/14E, 2 September 1999, The Hague, Health Council of the Netherlands.

Hettne, B. 2008, Introduction, in Hettne, B. (ed.), *Sustainable Development in a Globalized World*, Studies in Development, Security and Culture, vol. 1, Basingstoke, Palgrave Macmillan, 1–5.

Hewett, C. and Foley, J. 2000, *Plane Trading: Policies for Reducing the Climate Change Effects of International Aviation*, London, Institute for Public Policy Research.

Holden, E. and Linnerud, K. 2007, 'The sustainable development area: satisfying basic needs and safeguarding ecological sustainability', *Sustainable Development*, 15: 174–87.

Hooper, P., Heath, B. and Maughan, J. 2003, Environmental management and the air transport industry, in Upham, P., Maughan, J., Raper, D.W. and Thomas, C.S. (eds), *Towards Sustainable Aviation*, London, Earthscan, 115–30.

Horne, J.A., Pankhurst, F.L., Reyner, L.A., Hume, K.I. and Diamond, I.D. 1994, 'A field study of sleep disturbance: effects of aircraft noise and other factors on 5742 nights of actimetrically monitored sleep in a large complex sample', *Sleep*, 17(2): 146–59.

Horton, G. 2006, *Forecasts of CO_2 Emissions from Civil Aircraft for IPCC*, November 2006, Farnborough, QinetiQ.

Houghton, J. 2009, *Global Warming: The Complete Briefing*, 4th edition, Cambridge, Cambridge University Press.

Hubbard, H.H., Maglieri, D.J. and Copei, W.L. 1967, 'Research approaches to alleviation of airport community noise', *Journal of Sound and Vibration*, 5(2): 377–86.

Hume, K.I., Gregg, M., Thomas, C. and Terranova, D. 2003, 'Complaints caused by airport operations: an assessment of annoyance by noise level and time of day', *Journal of Air Transport Management*, 9: 153–60.

Hume, K.I., Van, F. and Watson, A. 1998, 'A field study of age and gender differences in habitual adult sleep', *Journal of Sleep Research*, 7: 85–94.

Hume, K.I. and Watson, A. 2003, The human health impacts of aviation, in Upham, P., Maughan, J., Raper, D.W. and Thomas, C.S. (eds), *Towards Sustainable Aviation*, London, Earthscan, 48–76.

Humphreys, I. 2003, Organizational and growth trends in air transport, in Upham, P., Maughan, J., Raper, D.W. and Thomas, C.S. (eds), *Towards Sustainable Aviation*, London, Earthscan, 19–35.

Hutton, T.J., McEnaney, B. and Crelling, J.C. 1999, 'Structural studies of wear debris from carbon-carbon composite aircraft brakes', *Carbon*, 37: 907–16.

IATA (International Air Transport Association) 1996, *Environmental Review 1996*, Montreal, IATA.

ICAO (International Civil Aviation Organization) 1981, *International Standards and Recommended Practices – Environmental Protection: Annex 16 to the Convention on International Civil Aviation, Volume II: Aircraft Engine Emissions*, Montreal, ICAO.

ICAO 1993, *International Standards and Recommended Practices – Environmental Protection: Annex 16 to the Convention on International Civil Aviation, Volume II: Aircraft Engine Emissions*, 2nd edition, Montreal, ICAO.

ICAO 1995, *ICAO Engine Exhaust Emissions Databank*, Montreal, ICAO.

ICAO 1999, ICAO adopts new aircraft engine emissions and noise standards, ICAO News Release, PIO 02/99, Montreal, ICAO.

ICAO 2001, *Aviation and Sustainable Development*, Background Paper No. 9, DESA/DSD/2001/9, prepared by the International Civil Aviation Organization, UN Department of Economic and Social Affairs, Commission on Sustainable Development, Ninth Session, 16–27 April 2001, New York.

ICAO 2002, *Statement by the International Civil Aviation Organization (ICAO) to the World Summit on Sustainable Development*, Johannesburg, South Africa, 26 August – 4 September 2002, Montreal, ICAO.

ICAO 2004, 'ICAO circular examines ways to minimize aircraft fuel use and reduce emissions', *ICAO Journal*, 59 (2): 23–4, 30.

ICAO 2005, *International Standards and Recommended Practices – Environmental Protection: Annex 16 to the Convention on International Civil Aviation, Volume I: Aircraft Noise*, 4th edition, Montreal, ICAO.

ICAO 2006a, *Aircraft noise* [online], available at http://www.icao.int/cgi/goto_m_atb.pl?/icao/en/env/noise.htm, accessed 11 May 2009.

ICAO 2006b, *Convention on Civil Aviation*, 9th edition, Doc. 7300/9, Montreal, ICAO.

ICAO 2007, *Statement by the Secretary General of the International Civil Aviation Organization (ICAO), Dr Taïeb Chérif, to the United Nations High-Level Event on Climate Change*, 24 September 2007 [online], available at http://www.un.org/webcast/climatechange/highlevel/2007/pdfs/ICAO-eng.pdf, accessed 17 November 2008.

Ignaccolo, M. 2000, 'Environmental capacity: noise pollution at Catania-Fontanarossa international airport', *Journal of Air Transport Management*, 6: 191–99.

IPCC (Intergovernmental Panel on Climate Change) 1992, *Climate Change 1992: The Supplementary Report to the IPCC Scientific Assessment*, (ed.) Houghton, J.T., Callander, B.A. and Varney, S.K., Cambridge, Cambridge University Press.

IPCC 1995, *Climate Change 1994, Radiative Forcing of Climate Change and an Evaluation of the IPCC IS92 Emission Scenarios*, Reports of Working Groups I and III of the Intergovernmental Panel on Climate Change, forming part of the IPCC Special Report to the First Session of the Conference of the Parties to the UN Framework Convention on Climate Change, Houghton, J.T., Meira, L.G., Filho, J., Bruce, H., Lee, B.A., Callander, E., Haites, E., Harris, N. and Maskell, K. (eds), Cambridge, Cambridge University Press.

IPCC 1999, *Aviation and the Global Atmosphere*, A Special Report of IPCC Working Groups I and III in Collaboration with the Scientific Assessment Panel to the Montreal Protocol on Substances that Deplete the Ozone Layer, (ed.) J.E. Penner, D.H. Lister, D.J. Griggs, D.J. Dokken and M. McFarland, Cambridge, Cambridge University Press.

IPCC 2000, *Emissions Scenarios: A Special Report of IPCC Working Group III: Summary for Policymakers*, Cambridge, Cambridge University Press.

IPCC 2001, *Climate Change 2001: Synthesis Report: Summary for Policymakers*, Cambridge, Cambridge University Press.

IPCC 2007, *Climate Change 2007: Synthesis Report*, Cambridge, Cambridge University Press.

Ising, H., Babisch, W. and Kruppa, B. 1999, 'Noise induced endocrine effects and cardiovascular risk', *Noise and Health*, 4: 37–48.

Jacobs, M. 1991, *The Green Economy: Environment, Sustainable Development and the Politics of the Future*, London, Pluto Press.

Jacobs, M. 1995, 'Sustainable development, capital substitution and economic humility: a response to Beckerman', *Environmental Values*, 4: 57–68.

Janelle, D.G. and Beuthe, M. 1997, 'Globalization and research issues in transportation', *Journal of Transport Geography*, 5(3): 199–206.

Janić, M. 1999, 'Aviation and externalities: the accomplishments and problems', *Transportation Research Part D*, 4: 159–80.

Janić, M. 2008, 'The potential of liquid hydrogen for the future 'carbon-neutral' air transport system', *Transportation Research Part D: Transport and Environment*, 13(7): 428–35.

Jardine, C. 2005, *Calculating the Environmental Impact of Aviation Emissions*, Report commissioned for Climate Care, June 2005, Oxford, Environmental Change Institute.

Jin, F., Wang, F. and Liu, Y. 2004, 'Geographic patterns of air passenger transport in China 1980–1998: imprints of economic growth, regional inequality, and network development', *The Professional Geographer*, 56: 471–87.

Job, R.F.S. 1996, 'The influence of subjective reactions to noise on health effects of the noise', *Environment International*, 22(1): 93–104.

Johnston, H.S. 1971, 'Reduction of stratospheric ozone by nitrogen dioxide catalysts from supersonic transport exhaust', *Science*, 173: 517–22.

Karmali, A. and Harris, M. 2004, 'ICAO exploring development of a trading scheme for emissions from aviation', *ICAO Journal*, 59(5): 11–13, 25.

Kesgin, U. 2006, 'Aircraft emissions at Turkish airports', *Energy*, 31: 372–384.

Kemp, D.D. 2004, *Exploring Environmental Issues: An Integrated Approach*, London, Routledge.

Kim, B.Y., Fleming, G.G., Lee, J.J., Waitz, I.A., Clarke, J.-P., Balasubramanian, S., Malwitz, A., Klima, K., Locke, M., Holsclaw, C.A., Maurice, L.Q. and Gupta, M.L. 2007, 'System for assessing Aviation's Global Emissions (SAGE), Part 1: model description and inventory results', *Transportation Research Part D: Transport and Environment*, 12(5): 325–46.

King, J.M.C. 2007, 'The Airbus 380 and Boeing 787: a role in the recovery of the airline transport market', *Journal of Air Transport Management*, 13: 16–22.

Ko, N.W.M. and Lei, P.C.K. 1982, 'Subjective responses of Chinese to aircraft noise', *Applied Acoustics*, 15(4): 251–61.

Koppert, A.J. 1996, 'New land-use planning guidance material would review possible solutions to problem of noise', *ICAO Journal*, 51(2): 21, 25.

Kryter, K.D. 1967, 'Acceptability of aircraft noise', *Journal of Sound and Vibration*, 5(2): 364–9.

Lambert, C. 2008, *Alternative Aviation Fuels, SBAC Aviation and Environment Briefing Papers, No. 4* [online], http://www.sbac.co.uk/pages/92567080.asp, accessed 12 June 2008.

Lash, S. and Urry, J. 1994, *Economies of Signs and Space*, London, Sage Publications.

Lauber, A. 1967, 'New airport monitoring system for Zürich airport', *Journal of Sound and Vibration*, 5(2): 344–6.

Lee, D.S. 2004, The impact of aviation on climate, in Hester, R.E. and Harrison, R.M. (eds), *Transport and the Environment*, Issues in Environmental Science and Technology, No. 2, Cambridge, The Royal Society of Chemistry, 1–23.

Lee, D.S. 2009, Aviation and climate change: the science, in Gössling, S. and Upham, P. (eds), *Climate Change and Aviation: Issues, Challenges and Solutions*, London, Earthscan: 27–67.

Lee, D.S., Fahey, D.W., Forster, P.M., Newton, P.J., Wit, R.C.N., Lim, L.L., Owen, B. and Sausen, R. 2009, 'Aviation and global climate change in the 21st century', *Atmospheric Environment*, 43(22–23): 3520–37.

Lee, D.S., Owen, B., Graham, A., Fichter, C., Lim, L.L. and Dimitriu, D. 2005, *Allocation of International Emissions from Scheduled Air Traffic – Present Day and Historical*, Final Report to DEFRA Global Atmosphere Division, Manchester, Centre for Air Transport and the Environment, Manchester Metropolitan University.

Lee, D.S. and Raper, D.W. 2003, The global atmospheric impacts of aviation, in Upham, P., Maughan, J.A., Raper, D.W. and Thomas, C.S. (eds), *Towards Sustainable Aviation*, London, Earthscan, 77–96.

Lee, D.S. and Sausen, R. 2000, 'New directions: assessing the real impact of CO_2 emissions trading by the aviation industry', *Atmospheric Environment*, 34: 5337–8.

Lee, J.J., Waitz, I.A., Kim, B.Y., Fleming, G.G., Maurice, L. and Holsclaw, C.A. 2007, 'System for assessing Aviation's Global Emissions (SAGE), Part 2: uncertainty assessment', *Transportation Research Part D: Transport and Environment*, 12(6): 381–95.

Lei, Z. 2008, China, in Graham, A., Papatheodorou, A. and Forsyth, P. (eds), *Aviation and Tourism: Implications for Leisure Travel*, London, Earthscan, 279–89.

Lenton, T.M., Loutre, M.-F., Williamson, M.S., Warren, R., Goodess, C.M., Swann, M., Cameron, D.R., Hankin, R., Marsh, R. and Shepherd, J.G. 2006, *Climate Change on the Millennial Timescale*, Tyndall Centre for Climate Change Research, Technical Report 41, February 2006, Norwich, Tyndall Centre for Climate Change Research.

Lu, C. 2009, Aviation and economic development: the implications of environmental costs on different airline business models and flight networks, in Gössling, S. and Upham, P. (eds), *Climate Change and Aviation: Issues, Challenges and Solutions*, London, Earthscan: 193–219.

Lu, C. and Morrell, P. 2006, 'Determination and applications of environmental costs at different sized airports – aircraft noise and engine emissions', *Transportation*, 33(1): 45–61.

Luz, G.A., Raspet, R. and Schomer, P.D. 1983, 'An analysis of community complaints to noise', *Journal of the Acoustical Society of America*, 105: 3336–40.

MacGregor, J. 2006, *Ecological space and a low-carbon future: crafting space for equitable economic development in Africa*, Fresh Insights Series No. 8, DfID/IIED/NRI [online], available at http://www.agrifoodstandards.org/, accessed 29 August 2007.

MacGregor, J. and Vorley, B. (eds) 2006, *Fair miles? Weighing environmental and social impacts of fresh produce exports from sub-Saharan Africa to the UK (summary)*, Fresh Insights Series No. 9, DfID/IIED/NRI [online], http://www.agrifoodstandards.org/, accessed 29 August 2007.

Macintosh, A. and Wallace, L. 2008, 'International aviation emissions to 2025: can emissions be stabilised without restricting demand?', *Energy Policy*, 37(1): 264–73.

Madas, M.A. and Zografos, K.G. 2008, 'Airport capacity vs. demand: mismatch or mismanagement?', *Transportation Research Part A: Policy and Practice*, 42(1): 203–26.

Mander, S.L., Bows, A., Anderson, K.L., Shackley, S., Agnolucci, P. and Ekins, P. 2008, 'The Tyndall decarbonisation scenarios – Part I: development of a backcasting methodology with stakeholder participation', *Energy Policy*, 36(10): 3754–63.

Mangiarotty, R.A. 1971, 'The reduction of aircraft engine fan-compressor noise using acoustic linings', *Journal of Sound and Vibration*, 18:4, 565–76.

Mannstein, H. and Schumann, U. 2005, 'Aircraft induced contrail cirrus over Europe', *Meteorologische Zeitschrift*, 14: 549–54.

Marín, A.G. 2006, 'Airport management: taxi planning', *Annals of Operations Research*, 143: 191–202.

Maris, E., Stallen, P.J., Vermunt, R. and Steensma, H. 2007, 'Noise within the social context: annoyance reduction through fair procedures', *Journal of the Acoustical Society of America*, 121(4): 2000–2010.

Marquart, S. and Mayer, B. 2002, 'Towards a reliable GCM estimation of contrail radiative forcing', *Geophysical Research Letters*, 29: 1179, doi:10.1029/2001GL014075.

Marquart, S., Ponater, M., Mager, F. and Sausen, R. 2003, 'Future development of contrail cover, optical depth and radiative forcing: impacts of increasing air traffic and climate change', *Journal of Climate*, 16: 2890–2904.

Marsh, G. 2007, 'Airbus takes on Boeing with reinforced plastic A350XWB', *Reinforced Plastics*, 51(11): 26–7, 29.

Marsh, G. 2008, 'Biofuels: aviation alternative?', *Renewable Energy Focus*, 9(4): 48–51.

Maschke, C., Breinl, S., Grimm, R. and Ising, H. 1993, The influence of nocturnal aircraft noise on sleep and on catecholamine secretion, in Ising, H. and Kruppa, B. (eds), *Noise and Disease*, Stuttgart, Gustav Fischer, 402–7.

Mason, K.J. 2007, 'Airframe manufacturers: which has the better view of the future?', *Journal of Air Transport Management*, 13: 9–15.

Mathiesen, B.V., Lund, H. and Nørgaard, P. 2008, 'Integrated transport and renewable energy systems', *Utilities Policy*, 16(2): 107–16.

Mato, R.R.A.M. and Mufuruki, T.S. 1999, 'Noise pollution associated with the operation of Dar es Salaam International Airport', *Journal of Transportation Research Part D*, 4: 81–9.

May, M. 2002, 'The growth of tourism and air travel in relation to ecological sustainability', *International Journal of Tourism Research*, 4: 147–50.

May, M. and Hill, S.B. 2006, 'Questioning airport expansion – a case study of Canberra International Airport', *Journal of Transport Geography*, 14(6): 437–50.

Mayor, K. and Tol, R.S.J. 2007, 'The impact of the UK aviation tax on carbon dioxide emissions and visitor numbers', *Transport Policy*, 14(6): 507–13.

Mayor, K. and Tol, R.S.J. 2010, 'Scenarios of carbon dioxide emissions from aviation', *Global Environmental Change*, 20: 65–73.

Mendes, L.M.Z and Santos, G. 2008, 'Using economic instruments to address emissions from air transport in the European Union', *Environment and Planning*, 40: 189–209.

Meyer, R., Mannstein, H., Meerkötter, R., Schumann, U. and Wendling, P. 2002, 'Regional radiative forcing by line-shaped contrails derived from satellite data', *Journal of Geophysical Research*, 107 (D10): 4104, doi:10.1029/2001JD000426.

Miller, J.D. 1974, 'Effects of noise on people', *Journal of the Acoustical Society of America*, 56(3): 729–64.

Miller, N.P. 2007, 'US National Parks and management of park soundscapes: a review', *Applied Acoustics*, 69(2): 77–92.

Miller, B. and Clarke, J.-P. 2007, 'The hidden value of air transportation infrastructure', *Technological Forecasting and Social Change*, 74: 18–35.

Minnis, P., Schumann, U., Doelling, D.R., Gierens, K. and Fahey, D.H. 1999, 'Global distribution of contrail radiative forcing', *Geophysical Research Letters*, 26: 1853–6.

Minnis, P., Ayers, J.K., Palikonda, R. and Phan, D. 2004, 'Contrails, cirrus trends, and climate', *Journal of Climate*, 17: 1671–85.

Morrell, P. and Lu, C. 2006, *The Environmental Cost Implications of Hub-Hub Versus Hub-Bypass Flight Networks*, Research Report 10, January 2006, Department of Air Transport, School of Engineering, Cranfield University.

Morrell, S., Taylor, R. and Lyle, D. 1997, 'A review of health effects of aircraft noise', *Australian and New Zealand Journal of Public Health*, 21(2): 221–36.

Mortimer, L.F. 1979, 'Aircraft engine emissions are under continuing surveillance', *Aircraft Engineering and Aerospace Technology*, 51(3): 6–8.

Moss, D., Warnaby, G., Sykes, S. and Thomas, C.S. 1997, 'Manchester Airport's Second Runway Campaign: The boundary spanning role of public relations in managing environmental organisational interaction', *Journal of Communication Management*, 2 (4): 320–34.

Myhre, G. and Stordal, F. 2001, 'On the tradeoff of the solar and infrared radiative impact of contrails', *Geophysical Research Letters*, 28: 3119–22.

Nero, G. and Black, J.A. 2000, 'A critical examination of an airport noise mitigation scheme and an aircraft noise charge: the case of capacity expansion and externalities at Sydney (Kingsford Smith) airport', *Transportation Research Part D*, 5: 433–61.

Netcen (National Environmental Technology Centre) 2006, *Air Quality in the UK: 2005*, August 2006, Didcot, Netcen.

Nicol, D.J. 1977, 'The influence of aircraft size on airline operating costs', *The International Journal of Management Science*, 6(1): 15–24.

Nilsson, J.H. 2009, 'Low-cost aviation', in Gössling, S. and Upham, P. (eds), *Climate Change and Aviation: Issues, Challenges and Solutions*, London, Earthscan: 113–29.

Njegovan, M. 2006, 'Are shocks to air passenger traffic permanent or transitory? Implications for long-term air passenger forecasts for the UK', *Journal of Transport Economics*, 40(2): 315–28.

Norgia, N. 1999, 'A graphical optimization of take-off noise abatement procedures for subsonic aircraft', *Journal of Sound and Vibration*, 222(3): 489–501.

Nygardis, M. 1999, 'Airports are under growing pressure to change as authorities respond to conflicting demands', *ICAO Journal*, 54(7): 4–6, 27.

O'Connell, J.F. 2008, India, in Graham, A., Papatheodorou, A. and Forsyth, P. (eds), *Aviation and Tourism: Implications for Leisure Travel*, London, Earthscan, 267–78.

O'Riordan, T. 1988, The politics of sustainability, in Turner, R.K. (ed.), *Sustainable Environmental Management: Principles and Practice*, London, Belhaven Press, 29–50.

OEF (Oxford Economic Forecasting) 1999, *The Contribution of the Aviation Industry to the UK Economy*, Final Report, November 1999, Oxford, OEF.

OEF 2002, *The Economic Contribution of the Aviation Industry to the UK: Part 2 – Assessment of Regional Impact*, May 2002, Oxford, OEF.

OEF 2006, *The Economic Contribution of the Aviation Industry in the UK*, October 2006, Oxford, OEF.

Ohrstrom, E. and Rylander, R. 1982, 'Sleep disturbance effects of traffic noise – a laboratory study on after effects', *Journal of Sound and Vibration*, 84(1): 87–103.

Oken, D. 2000, 'Multiaxial diagnosis and the psychosomatic model of disease', *Psychosomatic Medicine*, 62: 1739–48.

Owen, B. and Lee, D.S. 2005, *International Aviation Emissions Allocation – Allocation Options 2 and 3*, Manchester, Centre for Air Transport and the Environment, Manchester Metropolitan University.

Owen, B. and Lee, D.S. 2006, *Allocation of International Aviation Emissions from Scheduled Air Traffic – Future Cases, 2005 to 2050, – CPEG7*, Final Report to DEFRA Global Atmosphere Division, March 2006, Manchester, Centre for Air Transport and the Environment, Manchester Metropolitan University.

Pastowski, A. 2003, Climate policy for civil aviation: actors, policy instruments and the potential for emissions reductions, in *Towards Sustainable Aviation*, P. Upham, J. Maughan, D. Raper and C. Thomas, (eds), London, Earthscan, 179–95.

Pattarini, C.B. 1967, 'The aviation noise problem', *Journal of Sound and Vibration*, 5:2, 370–372.

Patterson, T., Bastianoni, S. and Simpson, M. 2006, 'Tourism and climate change: two-way street, or vicious/virtuous circle?', *Journal of Sustainable Tourism*, 14(4): 339–48.

Peace, H., Maughan, J., Owen, B. and Raper, D. 2006, 'Identifying the contribution of different airport related sources to local urban air quality', *Environmental Modelling and Software*, 21: 532–38.

Pearce, B. and Pearce, D. 2000, *Setting Environmental Taxes for Aircraft: A Case Study of the UK*, CSERGE Working Paper GEC 2000–26 [online], http://www.uea.ac.uk/env/cserge/pub/wp/gec/gec_2000_26.htm, accessed 11 June 2008.

Peeters, P., Gössling, S. and Becken, S. 2006, 'Innovation towards tourism sustainability: climate change and aviation', *International Journal of Innovation and Sustainable Development*, 1(3): 184–200.

Peeters, P. and Schouten, F. 2006, 'Reducing the ecological footprint of inbound tourism and transport to Amsterdam', *Journal of Sustainable Tourism*, 14(2): 157–71.

Peeters, P. and Williams, V. 2009, Calculating emissions and radiative forcing, in Gössling, S. and Upham, P. (eds) *Climate Change and Aviation: Issues, Challenges and Solutions*, London, Earthscan: 69–87.

Peeters, P., Williams, V. and de Haan, A. 2009, Technical and management reduction potentials, in Gössling, S. and Upham, P. (eds), *Climate Change and Aviation: Issues, Challenges and Solutions*, London, Earthscan: 293–307.

Pels, E. 2008, The Environmental Impacts of Increased International Air Transport: Past Trends and Future Perspectives, Paper presented to the Global Forum on Transport and Environment in a Globalising World, 10–12 November 2008, Guadalajara, Mexico.

Pepper, D. 1993, *Eco-Socialism: From Deep Ecology to Social Justice*, London, Routledge.

Pepper, D. 1996, *Modern Environmentalism: An Introduction*, London, Routledge.

Petzold, A. 2001, 'Aircraft engine exhaust measurement', *Air and Space Europe*, 3(1–2): 92–5.

Petzold, A., Döpelheuer, A., Brock, C.A. and Schröder, F. 1999, 'In situ observation and model calculation of black carbon emission by aircraft at cruise altitude', *Journal of Geophysical Research*, 104: 22171–81.

Ponater, M., Pechtl, S., Sausen, R., Schumann, U. and Hüttig, G. 2006, 'Potential of the cryoplane technology to reduce aircraft climate impact: a state-of-the-art assessment', *Atmospheric Environment*, 40: 6928–44.

Ponting, C. 2007, *A New Green History of the World: The Environment and the Collapse of Great Civilisations*, revised edition, London, Vintage.

Porritt, J. 2005, *Capitalism as if the World Matters*, London, Earthscan.

Porter, N., Flindell, I.H. and Berry, B.F. 1998, *Health Effect Based Noise Assessment Methods: A Review and Feasibility Study*, National Physics Laboratory, NPL Report CMAM 16, Teddington, NPL.

Porter, N., Kershaw, A. and Ollerhead, J.B. 2000, *Adverse Effects of Night-Time Aircraft Noise*, NATS R&D Report 9964, London, NATS.

Price, T. and Probert, D. 1995, 'Environmental impacts of air traffic', *Applied Energy*, 50: 133–62.

Raguraman K. 1995, 'The role of air transportation in tourism development: a case study of the Philippines and Thailand', *Transportation Quarterly*, 49: 113–24.

Ralph, M.O. and Newton, P.J. 1996, 'Experts consider operational measures as means to reduce emissions and their environmental impact', *ICAO Journal*, 51(2): 9–10.

Randles, S. and Mander, S. 2009, Practice(s) and ratchet(s): a sociological examination of frequent flying, in Gössling, S. and Upham, P. (eds) *Climate Change and Aviation: Issues, Challenges and Solutions*, London, Earthscan: 245–71.

RCEP 1994, *Eighteenth Report: Transport and the Environment*, Cm 2674, October 1994, London, RCEP.

RCEP 2002, *The Environmental Effects of Civil Aircraft in Flight*, Special Report, November 2002, London, RCEP.

Redclift, M. 1984, *Development and the Environmental Crisis: Red or Green Alternatives?*, London, Methuen.

Redclift, M. 1987, *Sustainable Development: Exploring the Contradictions*, London, Methuen.

Reijneveld, S.A. 1994, 'The impact of the Amsterdam aircraft disaster on reported annoyance by aircraft noise and on psychiatric disorders', *International Journal of Epidemiology*, 23(2): 333–40.

Rengaraju, V.R. and Arasan, V.T. 1992, 'Modeling for air travel demand', *Journal of Transportation Engineering*, 118(3): 371–80.

Revoredo, T.C. and Slama, J.G. 2008, 'Noise metrics comparison and its use on urban zoning in airport surveys: a Brazilian case study', *Journal of Air Transport Management*, 14(6): 304–7.

Rhoades, D.L. 2004, 'Sustainable development in African civil aviation: problems and policies', *International Journal of Technology, Policy and Management*, 4: 28.

Riddington, G., 2006, 'Long range air traffic forecasts for the UK: a critique', *Journal of Transport Economics and Policy*, 40(2): 297–314.

Roberts, J. 2004, *Environmental Policy*, Routledge Introductions to Environment Series, London, Routledge.

Rogers, H.L., Lee, D.S., Raper, D.W., Forster, P.M. de F., Wilson, C.W. and Newton, P. 2002, 'The impacts of aviation on the global atmosphere', *The Aeronautical Journal*, October 2002: 521–46.

Rolls-Royce 2007, *Market Outlook 2007: Forecast 2007–2026*, Derby, Rolls Royce plc.

Rousse, O. 2008, 'Environmental and economic benefits resulting from citizens' participation in CO_2 emissions trading: an efficient alternative solution to the voluntary compensation of CO_2 emissions', *Energy Policy*, vol. 35: 388–97.

Royal Academy of Engineering 2003, House of Commons Transport Select Committee: Inquiry into Aviation, Memorandum submitted by the Royal Academy of Engineering, April 2003, London, Royal Academy of Engineering.

Rypdal, K., Rive, N., Aström, S., Karvosenoja, N., Aunan, K., Bak, J.L., Kupiainen, K. and Kukkonen, J. 2007, 'Nordic air quality co-benefits from European post-2012 climate policies', *Energy Policy*, 35(12): 6309–22.

Sausen, R., Gierens, K., Ponater, M. and Schumann, U. 1998, 'A diagnostic study of the global distribution of contrails part I: present day climate', *Theoretical and Applied Climatology*, 61: 127–51.

Sausen, R., Isaksen, I., Grewe, V., Hauglustaine, D., Lee, D.S., Myhre, G., Köhler, M.O., Pitari, G., Schumann, U., Stordal, F. and Zerefos, C. 2005, 'Aviation radiative forcing in 2000: an update on IPCC (1999)', *Meteorologische Zeitschrift*, 14(4): 555–61.

Sawyer, F.L. 1967, 'Aircraft noise and the siting of a major airport', *Journal of Sound and Vibration*, 5(2): 355–63.

Schultz, T.J. 1978, 'Synthesis of social surveys on noise annoyance', *Journal of the Acoustical Society of America*, 64: 377–405.

Schumann, U. 1994, 'On the net effect of emissions from aircraft engines on the state of the atmosphere', *Annales Geophysicae*, 12: 365–84.

Schumann, U. 1996, 'On conditions for contrail formation from aircraft exhausts', *Meteorologische Zeitschrift*, 5: 4–23.

Schumann, U. and Wendling, P. 1990, Determination of contrails from satellite data and observational results, in Schumann, U. (ed.), *Air Traffic and the Environment – Background, Tendencies and Potential Global Atmospheric Effects*, Heidelberg, Springer-Verlag, 138–53.

Schürmann, G., Schäfer, K., Jahn, C., Hoffmann, H., Bauerfeind, M., Fleuti, E. and Rappenglück, B. 2007, 'The impact of NO_x, CO and VOC emissions on the air quality of Zurich airport', *Atmospheric Environment*, 41: 103–18.

Schuurman, F.J. 1993, *'Beyond the Impasse: New Directions in Development Theory*, London, Zed Books.

SDC (Sustainable Development Commission) 2001a, *Aviation and Sustainable Development*, 1 April 2001, London, SDC.

SDC 2001b, *Review 2001: Headlining Sustainable Development*, 1 November 2001, London, SDC.

SDC 2002, *Air Transport and Sustainable Development – A Submission from the SDC*, 1 November 2002, London, SDC.

SDC 2008, *Breaking the Holding Pattern: A New Approach to Aviation Policy Making in the UK*, London, SDC.

Sealy, K.R. 1966, *The Geography of Air Transport*, revised edition, London, Hutchinson University Library.

Seidel, S. and Rossell, M. 2001, 'Potential policy tools for reducing emissions shift emphasis to economic incentives', *ICAO Journal*, 56(4): 27–9, 34.

Seinfeld, J.H. and Pandis, S.N. 2006, *Atmospheric Chemistry and Physics: From Air Pollution to Climate Change*, 2nd edition, Hoboken, New Jersey, John Wiley and Sons, Inc.

Sekuler, R. and Blake, R. 1994, *Perception*, New York, McGraw-Hill.

Selye, H. 1956, *The Stress of Life*, New York, McGraw-Hill.

Sen, A. 1999, *Development as Freedom*, Oxford, Oxford University Press.

Shaw, S. and Thomas, C. 2006, 'Discussion note: social and cultural dimensions of air travel demand: hyper-mobility in the UK?', *Journal of Sustainable Tourism*, 14(2): 209–15.

Simmons, I.G. 1998, Towards an environmental history of Europe, in Butlin, R.A. and Dodgshon, R.A. (eds), *An Historical Geography of Europe*, Oxford, Oxford University Press, 335–61.

Simões, A.F. and Schaeffer, R. 2005, 'The Brazilian air transportation sector in the context of global climate change: CO_2 emissions and mitigation alternatives', *Energy Conversion and Management*, 46: 501–13.

Simpson, A. and Kent, R. 1999, 'Voluntary agreement enabled operators to provide self-made solutions to an environmental problem', *ICAO Journal*, 54(7): 23–4, 29–30.

Skogö, I. 2001, 'Public opposition to air transport development underscores importance of tackling noise issue', *ICAO Journal*, 56(4): 22–3.

Smith, M.J.T. 1992, 'Evolving noise issue could persist into the next century', *ICAO Journal*, 47(8): 11–13.

Smith, R.A. 2008, 'Enabling technologies for demand management: transport', *Energy Policy*, 36(12): 4444–8.

Somerville, H. 1993, The airline industry's perspective, in Bannister, D. and Button, K. (eds), *Transport, the Environment and Sustainable Development*, London, E & FN Spon, 161–74.

Spreng, M. 2000a, 'Central nervous system activation by noise', *Noise and Health*, 7: 49–57.

Spreng, M. 2000b, 'Possible health effects of noise induced cortisol increase', *Noise and Health*, 7: 59–63.

Stansfeld, S.A., Berglund, B., Clark, C., Lopez-Barrio, I., Fischer, P., Haines, M.M., Head, J., Hygge, W., van Kamp, I. and Berry, B.F. 2005, 'Aircraft and road traffic noise and children's cognition and health: a cross-national study', *The Lancet*, 365(9475): 1942–49.

Starke, L. 1990, *Signs of Hope: Working Towards Our Common Future*, Oxford, Oxford University Press.

Stedman, J.R., Kent, A.J., Grice, S., Bush, T.J. and Derwent, R.G. 2007, 'A consistent method for modelling PM_{10} and $PM_{2.5}$ concentrations across the United Kingdom in 2004 for air quality assessment', *Atmospheric Environment*, 41: 161–72.

Stern, N. 2007, *The Economics of Climate Change: The Stern Review*, Cambridge, Cambridge University Press.

Stockbridge, H.C.W and Lee, M. 1973, 'The psychosocial consequences of aircraft noise', *Applied Ergonomics*, 4: 44–5.

Stordal, F., Myhre, G., Stordal, E.J.G., Rossow, W.B., Lee, D.S., Arlander, D.W. and Svendby, T. 2005, 'Is there a trend in cirrus cloud cover due to aircraft traffic?', *Atmospheric Chemistry and Physics*, 5: 2155–62.

Stratford, A. 1974, *Airports and the Environment: A Study of Air-Transport Development and its Impact upon the Social and Economic Well Being of the Community*, Bristol, Macmillan Press.

Stuber, N., Forster, P., Rädel, G. and Shine, K. 2006, 'The importance of the diurnal cycle of air traffic for contrail radiative forcing', *Nature*, 441: 864–7.

Stuber, N., Sausen, R. and Ponater, M. 2001, 'Stratosphere adjusted radiative forcing calculations in a comprehensive climate model', *Theoretical and Applied Climatology*, 68: 125–35.

Sustainable Aviation 2005, *A strategy towards sustainable development of UK aviation* [online], available at http://www.sustainableaviation.co.uk, accessed 19 April 2009.

Swan, W. 2007, 'Misunderstandings about airline growth', *Journal of Air Transport Management*, 13: 3–8.

T&E (European Federation for Transport and Environment) 2006, *Clearing the Air: The Myth and Reality of Aviation and Climate Change*, Brussels, T&E and CAN-Europe (Climate Action Network Europe).

Takeshita, T. and Yamaji, K. 2008, 'Important roles of Fischer-Tropsch synfuels in the global energy future', *Energy Policy*, 36(8): 2773–84.

Tarnopolsky, A., Watkins, G. and Hand, D.J. 1980, 'Aircraft noise and mental health: prevalence of individual symptoms', *Psychological Medicine*, 10: 683–98.

Thomas, C.S. and Lever, M. 2003, Aircraft noise, community relations and stakeholder involvement', in Upham, P., Maughan, J., Raper, D.W. and Thomas, C.S. (eds), *Towards Sustainable Aviation*, London, Earthscan, 97–112.

Thomas, C.S. and Raper, D.W. 2000, 'The role of aero engineering in the sustainable development of the aviation industry', *The Aeronautical Journal*, 1037: 331–3.

Timms, P., Kelly, C. and Hodgson, F. 2005, *World Transport Scenarios Project*, Tyndall Centre Technical Report 25, April 2005, Norwich, Tyndall Centre for Climate Change Research.

Tomkins, J., Topham, N., Twomey, J. and Ward, R. 1998, 'Noise versus access: the impact of an airport in an urban property market', *Urban Studies*, 35(2): 243–58.

UK Air Quality Archive 2008, Air pollution [online, available at http://www.airquality.co.uk/what_causes.php, accessed 23 May 2009.

UK Government 2005, *Securing the Future: The UK Government Sustainable Development Strategy, Presented to Parliament by the Secretary of State for Environment, Food and Rural Affairs by Command of Her Majesty*, March 2005, Cm 6467, London, HMSO.

UN (United Nations) 1992, *United Nations Framework Convention on Climate Change*, New York, UN.

UN 1998, *Kyoto Protocol to the United Nations Framework Convention on Climate Change*, New York, UN.

Unal, A., Hu, Y., Chang, M.E., Odman, M.T. and Russell, A.G. 2005, 'Airport related emissions and impacts on air quality: application to the Atlanta International Airport', *Atmospheric Environment*, 39: 5787–98.

UNCSD (United Nations Commission on Sustainable Development) 2001, *Report of the Inter-sessional Ad Hoc Working Group on Transport and Atmosphere, New York, 6–9 March 2001*, E/CN/17/2001/16, New York, UNCSD.

UNCTAD (United Nations Conference on Trade and Development) 1999a, *Air Transport Services: The Positive Agenda for Developing Countries*. Report by the UNCTAD Secretariat, TD/B/COM.1/EM.9/2, 16 April 1999, Geneva, UNCTAD.

UNCTAD 1999b, Clarifying issues on air transport services to define the elements of the positive agenda of developing countries as regards both the GATS and specific sector negotiations of interest to them: agreed conclusions, TD/B/COM.1/EM.9/L.1, 25 June 1999, Geneva, UNCTAD.

UNCTAD 1999c, *Report of the Expert Meeting on Air Transport Services: Clarifying Issues to Define the Elements of the Positive Agenda of Developing Countries as Regards both the GATS and Specific Sector Negotiations of Interest to Them*, Held at the Palais des Nations, Geneva, 21–23 June 1999, TD/B/COM.1/25. TD/B/COM/EM.9/3, 23 August 1999, Geneva UNCTAD.

Underwood, B.Y., Brightwell, S.M., Peirce, M.J. and Walker, C.T. 2001, *Air Quality at UK Regional Airports in 2005 and 2010: A Report Produced for DETR*, AEAT/ENV/R/0453, February 2001, Warrington, AEA Technology plc.

Underwood, B.Y., Walker, C.T. and Mackenzie, J. 1996, *Air Quality Implications of Heathrow T5: Calculated PM_{10} Emissions and Concentrations*, Didcot, AEA Technology plc.

UNDP (United Nations Development Programme) 2006, *Human Development Report 2006: Beyond Scarcity: Power, Poverty and the Global Water Crisis*, New York, UNDP.

UNDP 2007, *Human Development Report 2007/2008: Fighting Climate Change: Human Solidarity in a Divided World*, New York, UNDP.

Unique (Flughafen Zürich AG) 2004, *Aircraft NO_x-Emissions within the Operational LTO Cycle*, August 2004, Zurich, Unique.

Unique 2005, *Engine Thrust Reverser Emissions at Zurich Airport*, January 2005, Zurich, Unique.

UNWTO (United Nations World Tourism Organization) 2007, '*International tourists 1995–2006 (millions)*' [online], http://www.world-tourism.org, accessed 14 April 2007.

UNWTO 2008, *Tourism Highlights: 2008 Edition* [online], http://www.unwto.org/facts/eng/highlights.htm, accessed 22 May 2009.

Upham, P. 2001, 'Environmental capacity of aviation: theoretical issues and basic research directions', *Journal of Environmental Planning and Management*, 44(5): 721–34.

Upham, P. 2003, Introduction: perspectives on sustainability and aviation, in Upham, P., Maughan, J., Raper, D.W. and Thomas, C.S. (eds), *Towards Sustainable Aviation*, London, Earthscan, 3–18.

Upham, P. and Gössling, S. 2009, Conclusion, in Gössling, S. and Upham, P. (eds), *Climate Change and Aviation: Issues, Challenges and Solutions*, London, Earthscan: 373–5.

Upham, P., Maughan, J., Raper, D. and Thomas, C. 2003, *Towards Sustainable Aviation*, London, Earthscan Publications.

Upham, P., Raper, D.W., Thomas, C.S., McLellan, M., Lever, M. and Lieuwen, A. 2004, 'Environmental capacity and European air transport: stakeholder opinion and implications for modelling', *Journal of Air Transport Management*, 10: 199–205.

Upham, P., Thomas, C.S., Gillingwater, D. and Raper, D.W. 2003, 'Environmental capacity and airport operations: current issues and future prospects', *Journal of Air Transport Management*, 9: 145–51.

Upham, P., Tomei, J. and Boucher, P. 2009, 'Biofuels, aviation and sustainability: prospects and limits', in Gössling, S. and Upham, P. (eds) *Climate Change and Aviation: Issues, Challenges and Solutions*, London, Earthscan: 309–28.

Vedantham, A. and Oppenheimer, M. 1998, 'Long-term scenarios for aviation: demand and emissions of CO_2 and NO_x', *Energy Policy*, 26(8): 625–41.

Vega, H. 2008, 'Air cargo, trade and transportation costs of perishables and exotics from South America', *Journal of Air Transport Management*, 14(6): 324–8.

Veitch, R. and Arkkelin, D. 1995, *Environmental Psychology: An Interdisciplinary Approach*, Englewood Cliffs, NJ, Prentice Hall.

Vincendon, M. and von Wrede, R. 1999, 'Aircraft builder continually seeking new ways to limit impact on the environment', *ICAO Journal*, 54(7): 20–22.

Viswanath, P.R. 2002, 'Aircraft viscous drag reduction using riblets', *Progress in Aerospace Sciences*, 38(6–7): 571–600.

Wahner, A., Geller, M.A., Arnold, F., Brune, W.H., Cariolle, D.A., Douglass, A.R., Johnson, C., Lister, D.H., Pyle, J.A., Ramaroson, R., Rind, D., Rohrer, F., Schumann, U. and Thomson, A.M. 1995, Subsonic and supersonic aircraft emissions, in World Meteorological Organization, *Scientific Assessment of Ozone Depletion: 1994*, World Meteorological Organization Global Ozone Research and Monitoring Project – Report No. 37, Geneva, World Meteorological Organization.

Walle, F. 1999, 'Demonstrated commitment to environmental protection is increasingly important for airlines', *ICAO Journal*, 54(7): 18–19, 29.

Wang, J. and Jin, F. 2007, 'China's air passenger transport: an analysis of recent trends', *Eurasian Geography and Economics*, 48: 469–80.

Wangler, Z.L. 2006, *Sub-Saharan African horticultural exports to the UK and climate change: a literature review*, Fresh Insights Series No. 2, DFID/IIED/NRI [online], available at http://www.agrifoodstandards. org, accessed 29 August 2007.

WCED (World Commission on Environment and Development) 1987, *Our Common Future*, Oxford, Oxford University Press.

Whitehead, C.J., Hume, K.I. and Muzet, A. 1998, 'Cardiovascular responses to aircraft noises in sleeping subjects', *Noise Effects*, 7: 141–4.

WHO (World Health Organization) 1999, *Guidelines for Community Noise*, Geneva, WHO.

Wilkinson, R.T. 1984, 'Disturbance of sleep by noise: individual differences', *Journal of Sound Vibration*, 95: 55–63.

Williams, A. 2007, *Comparative Study of Cut Roses for the British Market Produced in Kenya and the Netherlands*, Précis Report for World Flowers, 12 February 2007, Natural Resources Management Institute, Department of Natural Resources, Cranfield University.

Williams, A.G., Audsley, E. and Sandars, D.L. 2006, *Determining the Environmental Burdens and Resource Use in the Production of Agricultural and Horticultural Commodities*, Main Report, DEFRA Research Project IS0205, Natural Resources Management Institute, Department of Natural Resources, Cranfield University.

Williams, M.L. 2007, 'UK air quality in 2050 – synergies within climate change policies', *Environmental Science and Policy*, 10(2): 169–75.

Winther, M., Kousgaard, U. and Oxbøl, A. 2006, 'Calculation of odour emissions from aircraft engines at Copenhagen Airport', *Science of the Total Environment*, 366: 218–32.

Wit, R.C.N., Boon, B.H., van Velzen, A., Cames, M., Deuber, O. and Lee, D.S. 2005, *Giving Wings to Emissions Trading: Inclusion of Aviation Under the European Emission Trading System (ETS): Design and Impacts*, Report for the European Commission, DG Environment, Delft, CE.

Wit, R.C.N, Kampman, B. and Boon, B.H. 2004, *Climate Impacts from International Aviation and Shipping: State-of-the-Art on Climatic Impacts, Allocation and Mitigation Policies*, Report for the Netherlands Research Programme on Climate Change, Scientific Assessments and Policy Analysis (NRP-CC), Delft, CE.

Wohlfrom, K.H., Eichkorn, S., Arnold, F. and Schulte, P. 2000, 'Massive positive and negative ions in the wake of a jet aircraft: detection by a novel aircraft-based large ion mass spectrometer (LIOMAS)', *Geophysical Research Letters*, 27: 3853–6.

Woodcock, J., Banister, D., Edwards, P., Prentice, A.M. and Roberts, I. 2007, 'Energy and transport', *The Lancet*, 370(9592), 22–28 September 2007: 1078–88.

World Bank 2009, *World Development Report 2009: Reshaping Economic Geography*, Washington, DC, World Bank.

World Bank 2010, *World Development Report 2010: Development and Climate Change*, Washington, DC, World Bank.

WRI (World Resources Institute) 2005, *Navigating the Numbers*, Washington, DC, WRI.

Wright, P. 1991, 'Air passenger transport', *Impact of Science on Society*, 41(2): 191–200.

Wylie, D., Jackson, D.L., Menzel, W.P. and Bates, J.J. 2005, 'Trends in global cloud cover in two decades of HIRS observations', *Journal of Climate*, 18: 3021–31.

Yamin, F. and Depledge, J. 2004, *The International Climate Change Regime: A Guide to Rules, Institutions and Procedures*, Cambridge, Cambridge University Press.

Young, E.M. 1997, *World Hunger*, Routledge Introductions to Development Series, London, Routledge.

Yu, K.N., Cheung, Y.P., Cheung, T. and Henry, R.C. 2004, 'Identifying the impact of large urban airports on local air quality by nonparametric regression', *Atmospheric Environment*, 38: 4501–7.

Zerefos, C.S., Eleftheratos, K., Balis, D.S., Zanis, P., Tselioudis, G. and Meleti, C. 2003, 'Evidence of impact of aviation on cirrus cloud formation', *Atmospheric Chemistry and Physics*, 3: 1633–44.

Index

A

abatement costs 6, 7, 9, 82, 86, 115, 207
ACARE 117, 120, 208
acid rain 14
acoustic linings 11, 152
adaptation 44, 49, 54–55
adaptive structures 68, 110
additives 21, 70
AERO2k 45
aerodynamic 4, 8, 33, 67, 110
    diameter 28, 91, 97, 107
aerosol 28, 42, 44, 51–52, 56, 60, 64, 66, 95, 97, 99, 107
Africa 16, 173, 187, 198
agreement
    See also air service agreement; Kyoto Protocol
    climate change 48, 55, 77, 206
    international 6, 48, 55, 77, 178
    local 4, 148, 158, 162, 194
    regulatory 34, 156
    UNCED process 178
    voluntary 9, 10, 74, 86, 116, 122, 207, 209
air freight 1, 19, 30, 34, 36, 38, 40-41, 87, 182, 198
air passenger duty 80, 118
air pollution 3, 5, 12–14, 34, 47, 87–94, 100–108, 113–114, 120–121, 135, 194, 205, 208
air quality
    management 87–88, 92, 107–108, 208
    monitoring 31, 93
    objectives 88, 92–93, 96–98, 100, 109–110, 120, 122, 205, 209
    strategy 92
air service agreement 6, 79, 80, 196, 203

Air Transport Action Group (ATAG) 180–181
air traffic management (ATM)
    service providers 120, 146, 153, 160, 180
    systems and procedures 6–8, 37, 43, 46, 71–74, 88, 108–109, 113–115, 149, 153
aircraft
    accidents 13, 14, 135, 138
    incidents 13, 14
    maintenance 5, 8, 14, 71, 73, 113, 114, 126
    servicing 5, 14, 19, 104
    size 13, 36, 41, 42
airframe
    airframe-engine combination 102, 104, 149
    deterioration 29, 33
    improvement 4, 6–8, 50, 66–70, 79, 88, 108–111, 113, 123, 152, 162, 209–210
    manufacturer 120, 180
    noise due to 126, 141, 152–153
    service life of 8, 73, 114
airline 13, 81, 85, 100, 120, 146, 148–149, 156, 160, 180, 184, 189, 197–198, 210
    business model 187
    emissions allocation to 78
    fleet replacement 154, 156, 162
    fuel consumption 72
    load factor 36
    procedures 71, 113, 153
airport
    air quality 3, 27–28, 87–89, 92–122, 205, 208–209
    capacity 12, 46, 80, 162, 186–187, 189
    city 157

climate impacts 49–50, 55, 66,
    72–73, 85
development 4, 89, 98, 100, 105,
    110, 135, 13 –139, 150–151,
    161, 189, 205, 209
expansion 6, 36, 87, 96, 159, 180
hub 41–42, 186
infrastructure 1, 5, 13, 14, 43, 69,
    112, 120–125, 131, 139–140,
    151, 161, 209
local environmental impacts 3, 5,
    10, 14, 19, 30–31, 47, 121, 123,
    133, 182, 186
management 12
noise 3, 9, 123–142, 146–162,
    205, 209–210
peripheral 41
policy 81–82, 84–85, 197–198,
    209
supply chains 184
surface transport 5, 14, 30, 37,
    66, 87, 97–100, 104–105, 111,
    118–119, 135
airspace 7, 34, 37, 43, 46, 71-72, 113,
    151, 153, 186
albedo 52
ALFA aircraft plume analysis facility
    31, 95, 106
altitude 4, 23, 32–33, 59, 63, 68, 72,
    106, 111, 147, 154
Amazonas 198
annoyance 123–126, 132–140,
    143–146, 161, 209
anxiety 123, 134, 136, 138, 141,
    144–145, 209
approach (phase of flight) 8, 72–73,
    104, 115, 136, 141, 150, 153
APU – *see* auxiliary power unit
arrival
    See also arrival management
    country of 77–78
    routes 3, 87, 134, 149–150, 159
arrival management 8, 73, 115
Asia 3, 16, 38, 41, 124, 186–187
auxiliary power unit (APU) 9, 73, 97,
    104, 111, 115, 126, 154

B
behavioural change 9, 74, 116
benzene 91, 95, 99
biofuel 8, 30, 69–70, 112
blended-wing body 8, 68, 110
boundary layer 100, 107, 113, 154
business jet 4, 41
Brazil 69, 198–199

C
CAA – *see* Civil Aviation Authority
capacity
    airspace 71
    constraints 37, 41, 46, 80, 158,
        187, 189
    environmental 6, 38, 46, 175, 186
    of the air transport system 37,
        184–186, 189
carbon
    See also carbon dioxide, carbon
        fibre, carbon monoxide
    budget 4, 24, 34, 48, 206
    elemental 21, 22, 27–29, 97
    footprint 15, 66
    intensity 44, 192
    leakage 80
    low-carbon 55, 84
    management 15
    neutrality 10, 15, 84–86, 207
    offsetting 7, 9–10, 16, 50, 74, 84,
        86, 207
    price 55, 79
    reduction target 195
    reservoir 56
    tax 79–80
carbon dioxide 3, 21, 22, 24, 47, 50,
    52, 57, 65, 91, 152, 205 *See
    also* carbon dioxide equivalent
carbon dioxide equivalent 84
carbon fibre 68, 110
carbon monoxide 12, 21, 22, 29, 47,
    90, 95
cargo 14, 36, 38, 72, 78, 114
catalytic converter 30
certification 12, 30, 33–34, 75, 88,
    102, 108, 116–117, 121, 154,
    156, 208

charges
  emissions 7, 9, 11, 74, 78– 81, 86,
    93, 116, 118, 121, 122, 207, 209
  noise-related 160
chemical transport model 32
chemi-ion 29
Chicago Convention 12, 79, 155
China 16, 38, 41, 173, 184
cirrus 4, 5, 13, 23, 28, 56, 57, 60–65,
    68, 85, 111
city-pair 42, 187
Civil Aviation Authority (CAA) 178
climate
  See also climate change
  policy 10, 50, 57, 66, 73–85, 125,
    207
  radiative forcing of 3, 4, 15, 47,
    52–53, 62–64, 85
  response 52, 54, 62, 64
  sensitivity 13, 52
  system 24, 50–55, 57, 64
  variability 51
climate change 1, 3–6, 10–13, 15, 20,
    44, 47, 49–86, 152, 203, 205,
    207–208, 210, 211
  adaptation 54
  agreement 48, 49, 55, 77, 194, 206
  and sustainable development 6,
    167, 169, 173–174, 194, 202
  anthropogenic 51–52
  costs 184
  mitigation 11, 54, 65, 70, 74, 76,
    194, 200, 207
climb (phase of flight) 104, 154 *See
    also* expedited climb
cloud 52, 53, 60, 62 *See also* cirrus,
    cloud condensation nuclei
cloud condensation nuclei 28, 57, 60,
    62
code of conduct 9, 74
combustion 20–22, 26–30, 47, 52,
    56, 58–59, 66, 68, 70, 88–89,
    96–99, 105, 111–113
  products 21–22, 30
combustor 4, 21, 26–27, 58, 68, 96,
    110–111, 121

commitment 4, 10, 46, 58, 76, 84, 86,
    92, 175–176, 178, 180–181, 207
Committee on Aircraft Engine
    Emissions (CAEE) 12–13
Committee on Aircraft Environmental
    Protection (CAEP) 13, 34, 42,
    75–76, 117, 119
Committee on Aircraft Noise (CAN)
    11, 13
communication 2, 39, 43, 123, 125,
    131, 134, 136, 137, 146, 192,
    195, 209
communications, navigation and
    surveillance (CNS) 73, 114–115
Community Noise Equivalent Level
    (CNEL) 128, 129, 133
community relations 140, 150, 161
competitiveness 9, 74, 79, 116, 184
computational fluid dynamics 106
condensation trail – *see* contrail
congestion 1, 71, 20, 113, 114, 135,
    184, 186
connectivity 1, 184, 196
consolidation model 41–42
contact – social 1
contamination (of land) 3, 5, 10
continuous descent approach (CDA)
    8, 73, 115, 153
contrail (condensation trail) 4, 5, 13,
    23, 27, 28, 56, 57, 60–65, 67,
    68, 70–72, 85, 111
control and handling system 8
control surface 68, 110
collaborative decision making (CDM)
    73
core cities 184
corporate responsibility 84, 120
cruise (phase of flight) 4, 12, 21, 22,
    75, 102, 106
cruising
  level 8, 9, 13, 23, 32, 47, 50, 59,
    63, 64, 67, 71, 73, 75, 113, 115
  speed 8, 71, 113

D
Day-evening-night average sound
    level (L*den*) 128, 129

Day-night average sound level (DNL
    or L*dn*) 129
decarbonise 195
deep ecology 171
de-greasing 11
de-icing 5, 11, 91
demand management 11, 45, 80, 86,
    118, 122, 194–196, 208, 209
departure
    See also departure management,
        noise abatement departure
        procedure
    country of 77–78
    phase of flight 141, 154
    procedures 9, 73, 115
    routes 87, 134, 149–150, 159
departure management 8, 73, 115
descent 8, 73, 113 *See also*
        continuous descent approach
        (CDA)
development
    See also Millennium Development
        Goal (MDG), sustainable
        development
    economic 2, 7, 44, 67, 77,
        155–156, 163–171, 182–190,
        195–199, 201, 205, 211
    human 194
    international 173, 190, 197, 198
    planning 174
    social 2, 44, 46, 67, 155, 163–166,
        169, 170, 189, 195, 197, 199,
        201, 204–205
    urban 159
diesel 66, 91, 97–100
dislocation 1
disturbance – *see* sleep disturbance
drag 7, 33, 67, 110, 126, 153 *See also*
        low-power/low-drag (LP/LD)
drainage 5, 14

E
eco-efficiency 170, 200
ecological
    effects 5, 91, 95, 172
    guidelines 202
    systems 172, 193

economic
    See also development
    activities 1, 175
    benefits 1–3, 6, 15, 135, 161,
        165–167, 176–177, 180–190,
        194, 199, 200, 205
    costs 1, 11, 161, 165–167, 176,
        177, 180, 182, 199, 205
    distortions 80
    driver 3, 39, 40, 155, 205
    growth 1, 2, 6, 7, 20, 38–44, 47,
        53, 156, 169–172, 178–184,
        193, 199–207, 211
    incentive 78, 81, 83, 118, 119
    restructuring 44, 167, 170–171
    sector 2, 176, 183, 185, 188, 191,
        197, 201
economies of scale 2, 182–184
ecosystem 54
education 2, 136, 145, 161, 171, 172,
    192
electricity 30, 90, 105 *See also* fixed
    electrical ground power
electroencephalograph 142
emission factor 20, 29–30, 75,
    102–105, 117
emissions
    See also emission factor, emissions
        trading scheme
    absolute 35, 47, 88, 108, 109, 121
    allocation of 77–78, 205
    charge 7, 9, 11, 80, 86, 93, 121,
        122, 207, 209
    greenhouse gas 1, 4, 10, 49,
        51–55, 66, 70, 76, 77, 86, 109,
        119, 194, 205, 207
    inventory 20, 32–33, 77, 88,
        93–94, 101–106
    limit 9, 11, 34, 74, 76–77, 86, 111,
        116–117, 122, 207, 209
    reduction 4, 8, 10, 30, 34, 49,
        66–74, 79–88, 93, 108–122,
        180, 205, 207–209
    scenarios 44, 53, 62, 65
    specific 34–35, 67, 71, 72, 75, 88,
        109–110, 114, 121
    target 10

tax 74, 93, 116, 118, 121
emissions trading scheme (ETS) 10,
  50, 65, 74, 82–83, 86, 116, 119,
  121–122, 207, 209 *See also*
  European Union Emissions
  Trading Scheme (EU ETS)
employment 2, 137, 161, 178, 180,
  182–185, 196–198
engine
  See also turbofan, turboprop
  efficiency 7, 26, 29, 67, 68, 75,
    79, 110–111, 117
  operating regimes 22–23
  performance 7, 8, 30, 31, 33, 47,
    67, 88, 110
  technology 6, 7, 8, 69, 70, 73–74,
    109–111, 115, 152, 162, 210
  testing 5, 14, 102, 126, 136, 141,
    154, 159
environmental
  See also environmental impact
    assessment
  capacity 6, 38, 46, 175, 186
  concern 6, 10, 11, 13, 83, 124,
    133, 165, 175, 199, 203
environmental impact assessment 151
equilibrium climate sensitivity 52
equity 6, 67, 78, 166, 168, 172–173,
  176, 190, 199, 210
Equivalent continuous sound level
  (L*eq*) 127–133, 138
erosion
  coastal 54
  soil 14
Europe 16, 91, 92, 97, 106, 124, 127,
  155, 156, 182, 186–187
European Commission 92
European Community 157
European Court of Human Rights 142
European Union (EU) 46, 82, 92,
  119–121, 156–158, 162, 177,
  178, 184, 187, 208, 210
  European Union Emissions
    Trading Scheme (EU ETS)
    10, 34, 82, 86, 207 *See also*
    ACARE, Single European Sky

exchange (cultural) 1, 2, 182
exhaust (engine) 11, 12, 20, 27–31, 60,
  95–96, 99, 102–103, 106–107,
  121
exhaustion 141, 144
expedited climb 9, 73, 115
export 2, 40, 183–185, 198

F
Federal Aviation Administration
  (FAA) 45, 102, 105, 126, 130
fire extinguishing 11
Fischer-Tropsch kerosene 70
fixed electrical ground power 9, 73,
  115, 153
flaps 153, 154
fleet 4, 12, 20, 26, 32, 34, 36, 40, 41,
  57, 60, 69, 70, 75, 79, 104, 112,
  117–118, 152, 155
  renewal 71, 81, 85, 113, 154, 156,
    162, 210
flight level 8, 72
forecast 9, 20, 26, 33, 37, 40–48, 58,
  60, 73, 75, 79, 96, 98, 109, 115,
  206 *See also* Forecasting and
  Economic Sub-Group (FESG)
Forecasting and Economic Sub-Group
  (FESG) 42–43, 45
foreign direct investment (FDA) 2,
  182, 183
fossil fuel 1, 29, 30, 47, 52, 56, 66,
  89–90, 98, 99, 111, 171, 181
fragmentation model 41–42
free radical 29
freight 1, 19, 30, 34, 36, 38, 40–41,
  87, 182, 198 *See also* freight
  tonne-kilometre
freight tonne-kilometre 34–36
fuel
  See also biofuel, combustor, fossil
    fuel, refuelling
  additives 70
  alternative 7, 66, 69, 81, 100,
    112–113, 121, 208
  availability 10, 43, 46, 69, 72, 86,
    112

cells 69, 112
consumption 1, 13, 29, 32–33, 56, 58, 61, 71–73, 103–105, 108, 114, 154, 181
  dumping 14
  international bunker 75–78
  low-sulphur 4, 8, 98
  optimisation 12
  price 43, 180
  spillage 5
  synthetic 69, 112
  tankering 12, 72–73, 80, 114
  tax 11, 45, 79–81, 118, 122, 207, 209
  unusable 8, 72, 114
  venting 12

G

gas 21, 23, 27–32, 47, 60, 64, 68, 89–91, 93, 95, 97–99, 103, 106, 111, 152 *See also* gas turbine, greenhouse gas
gas turbine 8, 20, 21
Geographic Information System (GIS) 94, 101, 139
Global Insight Forecasting Group 40
Global Warming Potential (GWP) 52, 64
globalisation 1–3, 20, 38–39, 47, 90, 160, 170, 188, 203, 206
go-around (missed approach) 136, 150
governance 7, 165, 170, 173–176, 179, 190–191, 194, 199, 200, 210
greenhouse effect 14, 24, 57, 59
greenhouse gas 4, 10, 13, 23, 24, 27, 42–59, 64, 66, 70, 76–77, 84, 86, 109, 193–194
gross domestic product (GDP) 2, 32, 33, 38–39, 43, 183–184, 197–198

H

habitat 3, 5, 10, 136, 175
health 54, 75, 90–92, 95–99, 121–124, 134–139, 142–147, 161, 174, 192, 196, 208, 209

holding 71–72, 113–114, 153
hub-bypass route planning 9, 73
hush-kit 152
hydraulic fluid 5
hydrocarbon 12, 20, 21–22, 27–28, 47, 90–91, 135, 208 *See also* unburned hydrocarbons
hydrofluorocarbons (HFCs) 52
hydrogen 8, 21, 27, 69, 112
hydroxyl 29
hypermobile 192
hypertension 144

I

ICAO 11–14, 33–34, 75–76, 82, 88, 102–104, 116–123, 154–157, 162, 177–178, 208 *See also* ICAO Engine Emissions Databank
ICAO Engine Emissions Databank 30, 33, 103–106
India 41
innovation 2, 6, 8, 69, 71, 75, 112, 117, 167, 183–184, 210
Instantaneous maximum sound level (L*max*) 127–131, 133, 140, 143–144
Institute for Public Policy Research (IPPR) 178–179
Integrated Noise Model (INM) 130
Intergovernmental Panel on Climate Change (IPCC) 15, 44, 51, 64
International Civil Aviation Organization – *see* ICAO
international tourist arrival (ITA) 3, 40

K

kerosene 8, 20–24, 27, 29, 30, 46, 50, 69–70, 73, 79, 80, 91, 104, 112, 115, 118, 180 *See also* Fischer-Tropsch kerosene, PROSENE
knowledge transfer 185
Kyoto Protocol 4, 24, 34, 48, 52, 55–56, 58, 75–77, 82, 206

L

laminar flow technology 68, 110
land contamination 3, 5, 10
landing 72, 81, 104, 107, 113, 130, 136, 141, 152, 160 *See also* landing gear, landing and take-off cycle (LTO)
landing gear 126
landing and take-off cycle (LTO) 23, 32–34, 56, 67, 75, 84, 87, 102–110, 113–116
Latin America 187
latitude 63–64
L*den* 128, 129
L*dn* 128, 129
lean-burn combustion 68, 111
legislation 4, 12, 31, 90, 92, 158, 162
leisure 1, 2, 123, 125, 134, 136, 182, 192, 195, 209
L*eq* 127–133, 138
levies 11, 74, 86, 116, 122, 207, 209
life cycle 19
lifestyle 1, 55, 135, 179, 192
Light Detection and Ranging (LIDAR) 93
limits
    capacity 37, 80, 158
    certification 75
    environmental 11, 12, 77, 86, 92, 93, 111, 120, 165, 169, 172, 179, 199, 207, 208
    operational 158
liquefied natural gas 69, 112
liquefied petroleum gas 98
livelihood 172, 202
L*max* 127–131, 133, 140, 143–144
load factor 8, 36, 41, 45
loading 7, 8, 71, 72, 88, 108, 113, 114
low-cost carrier (LCC) 1, 147, 187
low-power/low-drag (LP/LD) 8, 73, 115, 153
low-sulphur fuels 4, 8, 98

M

maintenance 5, 8, 14, 66, 71, 73, 92, 99, 113, 114, 126

market 2, 9, 16, 36, 40–43, 46, 67, 81, 83, 119, 156, 182–184, 187–190, 196–198, 201 *See also* market-based measure, market forecast, market maturity
market-based approach 7, 9, 50, 73–83, 86, 116, 118–119, 154, 160, 174, 203, 208
market forecast 20, 40–42, 45, 47, 206
market maturity 42–43
marketable permit – *see* tradable permit
manoeuvring 7, 8, 71, 84, 88, 107, 108, 113, 150
mass (aircraft) 34, 36, 72, 104, 114
mass flow 21, 22, 27
mass spectrometry 31
methane 4, 23, 27, 52, 57, 58
Middle East 16, 187
Millennium Development Goal (MDG) 173
missed approach 136, 150
mitigation 4, 44, 49, 54–55, 58, 70, 109, 148, 156, 157, 177, 191, 194, 207
modelling 16, 20, 32, 44, 93–94, 100, 118
    dispersion 88, 101, 106
    noise 130, 131, 148
    scenario 45
mobility 1, 46, 90, 177, 181, 193, 195, 204 *See also* hypermobile
monitoring 16, 20, 30–32, 92–93, 97, 100, 107, 118, 193
    noise 139, 140, 148–149
    physiological 136
multinational organisations 3, 39, 170, 188
multiplier effects 187, 197, 198

N

National Ambient Air-Quality Standards (NAAQS) 92
night 1, 61, 124, 127–131, 133, 136–138, 141–144, 147, 149, 154, 158, 160

nitric acid 29
nitric oxide 26, 58, 91, 93, 96
nitrogen 21, 22, 26, 58, 96 *See also*
    nitrogen dioxide, nitrogen
    oxides
nitrogen dioxide 26, 58, 91, 93, 95–96
nitrogen oxides 3, 21, 22, 26–27, 47,
    57–58, 87, 91, 95–96, 205
nitrous acid 29
nitrous oxide 52
noise 1, 4, 5, 8, 10–14, 42, 68, 111,
    121, 123–163, 181, 194, 200,
    204–205, 209–210 *See also*
    noise abatement departure
    procedure, noise preferential
    route (NPR)
  abatement 4, 8, 147, 149, 151,
    153–155
  airframe 126, 152–153
  ambient 126
  certification standards 154,
    156–157
  complaints 131, 134–135, 139–
    141, 147, 149–151
  contours 129–132, 152, 158–160
  exposure 126–132, 138–139, 143–
    145, 147–149, 153, 160–161,
    205, 209
  monitoring 139–140,
  modelling 130–131, 148
  nuisance 3, 123–125, 129, 131,
    133–136, 139–141, 147–151,
    154–162, 210
  subjective response to 13, 126,
    132–133, 137, 160
noise abatement departure procedure
    8, 154
noise preferential route (NPR) 8, 153
non-governmental organisation 156
North America 16, 124, 186, 187
North Atlantic Flight Corridor 71
Northern Hemisphere 16, 32, 59, 61,
    63

O
odour 28
oil 5, 40, 46, 69, 112

optical properties (of cloud) 61–62
oxygen 21, 22, 95, 99
ozone 4, 14, 23, , 27, 29, 56–58, 90,
    94–95
  stratospheric 14
  tropospheric 27, 56–58, 94–95

P
particles 3, 4, 13, 21–23, 27–33,
    47, 52–70, 85–98, 103–112,
    118–122, 205, 208–209
particulate matter 28, 91, 95, 97
payload 8, 72, 114
perfluorocarbons (PFCs) 52
petroleum 21
plume 12, 27–31, 95, 106–107, 121
polluter pays principle 160, 174, 179
polyaromatic hydrocarbons 91
poverty 169, 172–173, 193, 200 *See
    also* poverty reduction
poverty reduction 166, 168, 173, 176,
    190, 193–196, 200–201, 206,
    210–212
precautionary principle 174, 179
predict and provide approach 37,
    195
pressure 21, 26, 29, 68, 103, 111, 125
    See also pressure ration
pressure ratio 8, 21, 26, 68, 75, 111,
    117
productivity 1, 2, 183–184
projection 3, 33, 40, 45–46, 58, 93,
    100–101
PROSENE 69

Q
Quota Count 158

R
radiative forcing 3, 15, 47, 49, 52–53,
    56, 58–59, 61–65, 69, 70,
    75, 82–83 *See also* radiative
    forcing index
radiative forcing index (RFI) 64–65,
    82
recreation 1, 2
reduced thrust take-off 106, 154

refuelling 28, 66, 72, 80, 99, 101,
103, 105
regulators 83, 119, 139, 146, 153,
201, 203, 205
regulatory approaches 9, 74, 116
renewable energy sources 30, 108, 111
resources 46, 54, 161, 169, 172, 175,
178–179, 189, 193, 196, 198
revenue passenger-kilometre (RPK)
36, 38, 41, 43
revenue tonne-kilometre (RTK) 36
riblets 68, 110
route 8, 41–42, 71, 73, 80, 113, 130,
135, 148–150, 154, 156, 180,
187–188, 196–197 *See also*
hub-bypass route planning,
noise preferential route (NPR)

S
sanctions 74
scenario 4, 20, 24–26, 32, 42–48, 53,
58, 60, 62, 65, 79, 94, 110, 110,
121, 193, 206
serial complainer 139, 141, 150
servicing 5, 14, 19, 104
Single European Sky 71, 114
sleep disturbance 125, 130, 133–134,
136–138, 141–144, 146, 161
smog 90
smoke 12, 30, 33, 75, 89–90, 103, 116
See also smoke number
smoke number 103
social 1, 11, 15, 133, 165–168,
171–172, 175, 178, 189, 193,
198–200, 203–204, 210
benefits 2, 6, 15, 135, 161, 165,
167, 171, 176–182, 187, 190,
194, 197–200, 205
change 43, 44, 173, 179, 206
context 138, 167
costs 15, 17, 90, 160–161, 165,
167, 174–180, 182, 185, 193,
194, 197–199, 205
development 2, 44, 46, 67, 155,
163, 170–171, 185, 189,
195–199, 201, 204–205
groups 123, 125, 150–151, 161, 209

organisation 169–170, 173, 200,
204–205
responsibility 84, 120
services 196–197, 201
standards 170
surveys 170
transformations 2, 39
wellbeing 134
society 55, 167–168, 173, 177–179,
181, 190, 192–193, 195, 200,
202, 204
solar radiation 28, 52, 53, 57, 60–62
sonic boom 14
soot 4, 21–23, 28–29, 33, 47, 52, 56,
57, 60, 85, 95, 97
sound 124–138, 140, 143, 149, 152,
209 *See also* sound insulation,
sound exposure level (SEL)
sound insulation 131, 138, 152,
159–160
sound exposure level (SEL) 127–133
source apportionment 31–32, 88, 94,
101
South America 16
Southeast Asia 16, 38
Special Report on Emissions
Scenarios (SRES) 44, 46
specialisation 2, 182, 183
species (ecological) 54, 70, 176
speed 8, 29, 36, 43, 71, 72, 106, 113
speedbrakes 126, 153
stack – *see* holding
stratosphere 4, 12–14, 23, 26, 47, 60,
85
stress 54, 123, 134, 136–138, 141,
144–145, 209
subsidies 1, 7, 9, 11, 74, 78, 81, 86,
93, 116, 118–119, 121, 122,
207, 209
subsonic 12, 13, 26, 55–56, 59, 67
sulphate 4, 21, 23, 27–28, 47, 52, 56,
57, 60, 69, 85, 97, 112
sulphur 21, 22, 27–30, 44, 69, 89, 98,
112 *See also* sulphur dioxide,
sulphur oxides, sulphur trioxide,
sulphur hexafluoride

sulphur dioxide 21, 22, 27, 31, 89, 90,
    95, 98
sulphur hexafluoride 52
sulphur oxides 8, 21, 22, 27–28, 30,
    47, 57, 69, 95
sulphur trioxide 27
sulphuric acid 27
supersonic 4, 12, 14, 26, 55, 57, 60
surface transport 5, 19, 30, 37, 66,
    87, 97–102, 111, 113, 118–120,
    159, 205, 208
sustainable development 1, 6–10,
    15–17, 70, 84–85, 92, 112, 155,
    162–206, 210–211
System for assessing Aviation's Global
    Emissions (SAGE) 45

T
take-off 21, 102, 104, 107, 130, 136,
    141, 154, 158, 160 *See also*
    landing and take-off cycle
    (LTO), reduced thrust take-off
tankering – *see* fuel tankering
tax 2, 7, 9, 11, 45, 74, 78, 79–81, 83,
    86, 93, 116, 118, 121–122, 183,
    207
technological
    improvements 4, 7–8, 28, 32–34,
        43, 58, 67, 75, 88, 110–111,
        146, 152, 183, 200
    options 6–9, 50, 66–68, 74, 88,
        108–111, 115, 121–122, 125,
        152, 177, 207–209
temperature 21, 26, 29, 44, 50, 52–53,
    58–62, 68, 96, 111
time in mode 104–105
tolerance (of environmental impacts)
    4, 120, 124, 133, 147, 150, 154,
    208, 210
tourism 1–3, 6, 20, 38–40, 47, 90,
    184–185, 188, 192, 203, 206
tradable permit 7, 9, 74, 78, 81–82,
    93, 116, 118–119
trade 39, 81–82, 184–185 *See also*
    trade-off

trade-off 8, 16, 27, 55, 68–69, 83, 85,
    109, 111, 125, 152, 154, 208,
    210
TRADEOFF project 58, 63, 65, 72
tropopause 63
troposphere 4, 13, 23–24, 26, 47, 53,
    56, 59–60, 85
turbine 8, 20–21
turbofan 20–21, 99, 152
turbojet 12
turboprop 20
Tyndall Centre 46

U
unburned hydrocarbons 12, 27, 133
    See also hydrocarbon
undercarriage – *see* landing gear
unfair competition practices 196, 198,
    201
United Nations Conference on
    Environment and Development
    (UNCED) 165, 169, 178
United Nations Conference on the
    Human Environment (UNCHE)
    12
United Nations Framework
    Convention on Climate Change
    (UNFCCC) 51, 55, 75–78

V
Value Added Tax (VAT) 80–81, 118
vegetation 52, 66, 87, 89, 91, 99
venting – *see* fuel venting
volatile organic compounds (VOCs)
    95, 99, 103, 109
voluntary approaches 7, 9, 73–74,
    76, 83–84, 86, 109, 115, 116,
    119–121, 174, 203, 208
vortices 107

W
wake (aircraft) 31, 96, 98 *See also*
    wake turbulence
wake turbulence 150
waste 19, 24, 30
    disposal 5
    generation 3, 5, 10, 13, 14, 19

management 66, 99
water 14, 54
    consumption 3, 5, 10
    pollution 3, 5, 10, 13
    vapour 21, 22, 24, 26, 47, 53, 57,
        59–61, 65, 68, 70, 111
weight (aircraft) 7, 8, 21, 67–68,
        71–72, 110, 113–114, 152
wellbeing 123–125, 133–145, 147,
        157, 161, 170–172, 193, 198,
        200, 209, 212

wildlife 91, 136
wing-in-ground-effect design 68, 110
wing-tip device 68, 110
World Commission on Environment
        and Development (WCED)
        167, 169–170, 200, 211
World Summit on Sustainable
        Development (WSSD) 165, 169

Printed and bound by CPI Group (UK) Ltd, Croydon, CR0 4YY

24/10/2024

01779064-0003